RAISING MILK GOATS
THE MODERN WAY

RAISING

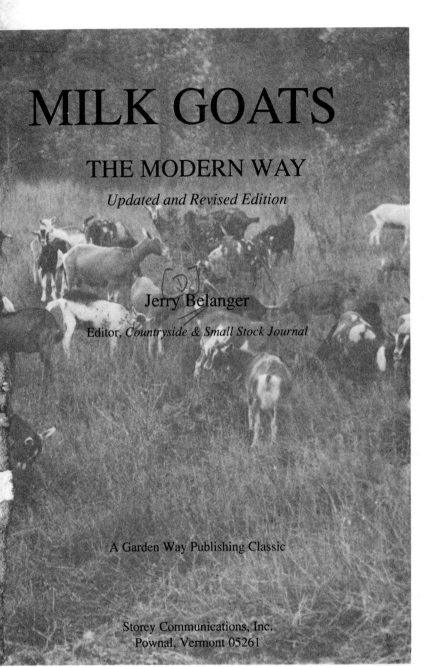

MILK GOATS

THE MODERN WAY

Updated and Revised Edition

Jerry Belanger

Editor, Countryside & Small Stock Journal

A Garden Way Publishing Classic

Storey Communications, Inc.
Pownal, Vermont 05261

Text design and production by Wanda Harper
Illustrations on pages 5-8, 10, 30, 42, 74, 75, 95, 117, 125, 128, 148, 168, and 178 by Brigita Furhmann

Storey Publishing books are available for special premium and promotional uses and for customized editions. For further information please call the Custom Publishing Department at 1-800-793-9396.

Printed in the United States by Capital City Press
Revised Edition
20   19   18   17   16   15

Garden Way Publishing was founded in 1973 as part of the Garden Way Incorporated Group of Companies, dedicated to bringing gardening information and equipment to as many people as possible. Today the name "Garden Way Publishing" is licensed to Storey Communications, Inc., in Pownal, Vermont. For a complete list of Garden Way Publishing titles call 1-800-827-8673. Garden Way Incorporated manufactures products in Troy, New York, under the TROY-BILT® brand including garden tillers, chipper/shredders, mulching mowers, sicklebar mowers, and tractors. For information on any Garden Way Incorporated product, please call 1-800-345-4454.

Library of Congress Cataloging-in-Publication Data
Belanger, Jerome D.
    Raising milk goats the modern way / Jerry Belanger. — Updated and rev. ed.
        p.   cm.   — (A Garden Way Publishing Classic)
    ISBN 0-88266-576-6 : $8.95
    1. Goats.   I. Title.
SF383   1989
636.3'9142—dc20

89-45737
CIP

Table of Contents

CHAPTER 1

Basic Information
about Goats

This book assumes that you are interested in goats and that you like these interesting and valuable animals, but it doesn't assume that you know anything about them. So let's start at the very beginning by looking at some basic facts and terms.

If you already know the basics, or if you're more interested in practical matters than in terminology and history, feel free to skip to the next chapter.

BASIC FACTS AND TERMS

Female goats are called does, or if they're less than a year old, sometimes doelings. Males are bucks, or bucklings. Young goats are kids. In polite goat company they are never "nannies" or "billies." Correct terminology is important to those who are working to improve the image of the dairy goat. People who think of a "nanny goat" as a stupid and smelly beast that produces small amounts of vile milk will at least have to stop to consider the truth if she's called a doe.

Does are *not* smelly. They are not mean, and of course they don't eat tin cans. They are dainty, fastidious about what they eat, intelligent (smarter than dogs, some scientists tell us), friendly, and a great deal of fun to have around.

Bucks have scent glands located between and just to the rear of the horns or horn knobs, and minor ones in other locations. They do smell, but the does think it's great and some goat raisers don't mind it either. The odor is worst during the breeding season, usually from September to about January. The scent glands can be removed, although some authorities frown on the practice.

But bucks have habits that make them less than ideal family pets even if they don't stink. For instance, they urinate all over their front legs and beards or faces, which tends to turn some people off.

In most cases the home dairy won't even have a buck, so we return to the fact that you can keep goats even if you have neighbors or if your barn is fairly close to the house and no one will be overpowered by goat aroma.

One of the problems with goat public relations is that everyone seems to have had one in the past or knows someone who did. Most of them were pets, and that's where the trouble lies.

A goat is not much bigger than a large dog (average weight for a doe is less than 150 pounds), it's no harder to handle, and it does make a good pet. But a goat is not a dog. People who treat it like one are asking for trouble, and when they get rid of the poor beast in disgust they bring trouble down on all goats and all goat lovers. If the goat "eats" the clothes off the line or nips off the rose bushes or the pine trees, strips the bark off young fruit trees or butts people, it's not the goat's fault but the owner's.

Would you let a cow or a pig roam free and then damn the whole species when one got into trouble? A goat is livestock. Would you condemn all dogs because one is vicious…after being chained, beaten, and teased? Children can have fun playing with goats, but when they "teach" young kids to butt people and that kid grows up to be a 200-pound male who still wants to play, there's bound to be trouble. Likewise, a mistreated animal of any species isn't likely to have a docile disposition.

The goat *(Capra hircus)* is related to the deer: not to dogs, cats, or even cows. It is a browser rather than a grazer, which means it would rather reach up than down for food. The goat also craves variety. Couple all that with its natural curiosity and nothing is safe from at least a trial taste.

Anything hanging, like clothes on a washline, is just too much for a goat's natural instincts to resist. Rose bushes and pine trees are high in vitamin C and goats love them. Leaves, branches, and the bark of young trees are a natural part of the goat's diet in the wild. If you treat a browsing goat like a carnivorous dog, of course you'll have problems! But don't blame the goat.

Like a cow or a pig, a goat needs a sturdily fenced pen. Each doe requires roughly 20 square feet of space. Goats do not require pasture, and unless it contains browse they probably won't utilize much of it anyway. They'll trample more than they eat. It's better to bring their food to them and feed them in a properly constructed manger, especially in a land- and labor-intensive small farm situation.

Goats are not lawn mowers. Most of them won't eat grass unless starved to it, and they won't produce milk on it.

Goats eat tin cans? Of course not. But they'll eat (or at least taste) the paper and glue on tin cans, which probably started the myth.

Never stake out a goat. There is too much danger of strangulation, and many goats have been injured or killed by dogs. Even the family pet you thought was a friend of the goat could turn on it.

All of this indicates that goats can be raised in a relatively small area. If there are no restrictive zoning regulations, they can be (and are) raised even on average-size lots in town.

Because goats are livestock, and more specifically dairy animals, they must be treated as such. That means not only proper housing and feed, but strict attention to regularity of care. If you can't or won't want to milk at 12-hour intervals—even if you're tired or under the weather — or if the thought of staying home weekends and vacations depresses you and you can't count on the help of a friend or neighbor, don't even start raising goats.

The rewards of goat raising are great and varied, but you don't get rewards without working for them.

A LITTLE HISTORY

Goats have been humanity's companions and benefactors throughout recorded history, and even before. There is evidence that goats were among the first, some say the first, animals to be domesticated by humans, perhaps as long as 10,000 years ago. They provided meat, milk, skins...and undoubtedly entertainment and companionship.

Wild goats originated in Persia and Asia Minor *(Capra aegagrus)*, the Mediterranean basin *(Capra prisca)*, and the Himalayas *(Capra falconeri)*. There were domesticated goats *(Capra hircus)* in Switzerland by the Early Stone Age. The first livestock registry in the world was organized in Switzerland in the 1600s—for goats.

Goats were distributed around the world by early explorers and voyagers. They were commonly carried on board ships as a source of milk and meat, and frequently ended up on shores far from home. Many became feral (returned to their wild nature) on deserted islands around the world. There were goats aboard the *Mayflower* on its famous voyage to America in 1620.

In Europe, goats provided more milk than cows did until well after the Middle Ages. Even today, in the world as a whole, it's said that more people use goat milk than cow milk. Goats are certainly more common in less fertile, or more arid, or "underdeveloped" countries than they are in the U.S. and Canada. They're more efficient animals than cattle are. Although they are more labor-intensive, this is of small concern in nonindustrialized countries and backyard dairies, and of no concern at all where there isn't enough feed for cattle to do well, or where a cow would produce more milk than a family can make good use of.

Goats are animals of the phylum *mammalia:* their young are born alive, and suckle on a secretion from the mammary glands, which, of course, is milk.

They are of the order *artiodactyla,* which means they are even-toed, hoofed mammals.

They belong to the family *bovidae,* which among other things means they are ruminants (with "four stomachs," like cows) and they have hollow horns, which they don't shed. (Some goats are naturally hornless, or polled. Many more are disbudded: the horn buds are burned out with a hot iron or with caustic before they start to grow. Some are dehorned: the horns are cut off after they grow. See Chapter 7.)

Goats belong to the genus *capra,* which includes only goats; we will discuss the species *hircus,* which is the domestic goat. Below that in the zoological scheme are breeds.

BREEDS OF GOATS

While all domestic goats have descended from a common parentage, there are many breeds, or subdivisions of the species, throughout the world. Only a few breeds are found in the United States.

These can be classed according to their main purpose: that is, meat, mohair, or milk. In this book we'll be concentrating on the goats that have been bred for milk production, although in most respects care is the same for all.

Bear in mind that many, perhaps most, American goats are not purebreds: they can't be identified as belonging to any particular breed. If these are fairly decent animals they're usually referred to as *grades;* if not, most people call them *scrubs.*

The most popular pure breed in America is the **Nubian.** Nubians can be any color or color pattern, but they're easily recognized by their long

Nubians are readily identified by their pendulous ears and Roman noses.

drooping ears and Roman noses.

It's commonly said that they originated in Africa, but technically, they made a stop along the way. Our Nubians are descendants of the Anglo-Nubian, which resulted from crossing native English goats with lop-eared breeds from Africa and India. The first three Nubians arrived in this country in 1909, imported by Dr. R. J. Gregg of Lakeside, California.

The Nubian is often compared with the Jersey of the cow world. The average Nubian will produce less milk than the average goat of any other breed, but the average butterfat content will be higher.

Averages can be misleading though. While the average production for a Nubian is said to be about 1,700 pounds in 305 days, with about 80 pounds of butterfat, there has been one record of 4,420 pounds of milk and 224.0 pounds of fat.

Next in popularity is the **Saanen** (pronounced *sah'-nen*). This is a pure white goat with erect ears and a "dished" face that is just the opposite of the Nubian's. Saanens originated in the Saane Valley of Switzerland and

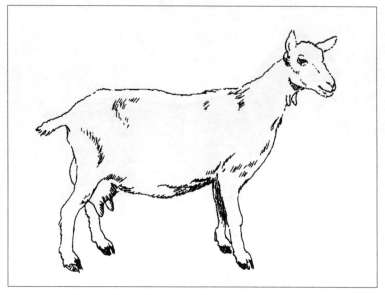

Saanens are always white and have "dish" or concave faces.

have enjoyed a wider distribution throughout the world than any other breed.

If Nubians can be compared to Jersey cattle, Saanens can be compared to Holsteins. They are large goats, with the highest average milk production: over 2,000 pounds in 305 days. Butterfat averages 3.5 percent on a yearly basis.

The first Saanens arrived in this country in 1904. Until recently, Saanens that were not a pure white or light cream color were discriminated against in purebred circles. Any that were colored or spotted could not be registered, and they were frequently disposed of.

That changed in the 1980s, when some Saanen breeders kept the colored or patterned animals, found that they were fine dairy animals, and started promoting them as a separate breed. They're not crossbreds: they're actually purebred Saanens, but with a "color defect" that results when both the sire and the dam carry a recessive color gene. Today these goats are called **Sables.** Many promoters of Sables like to refer to their goats as "The Saanens in party clothes."

French Alpines have erect ears and many of them have distinctive color patterns.

The **French Alpine** originated in the Alps and arrived in the U.S. in 1920, being imported by Dr. C. P. DeLangle. The color of Alpines varies greatly, ranging from white to black, and often with several colors and shades on the same animal.

There are recognized color patterns, such as the cou blanc (French for "white neck"). The white neck and shoulders shade through silver gray to a glossy black on the hindquarters. There are gray or black markings on the head. Another color pattern, the chamoisee, can be tan, red, bay, or brown, with black markings on the head, a black stripe down the back, and black stripes on the hind legs. The sundgau has black and white markings on the face and underneath the body. The pied is spotted or mottled; the cou clair has tan to white front quarters shading to gray with black hindquarters; and the cou noir has black front quarters and white hindquarters.

Alpines average about 2,000 pounds of milk with 3.5 percent butterfat.

You might also hear of **British Alpines, Rock Alpines** (named not because they like to climb on rocks any more than other goats do, but because they were developed in America by Mary Edna Rock), and **Swiss Alpines.**

Top: *The Oberhasli is a rich bay color with black stripes on the face, ears, belly, udder, and lower legs.* Bottom: *Toggenburgs have white markings on the face and rump.*

There are no more Swiss Alpines. No, they're not extinct: in 1978 their name was changed to **Oberhasli** (*oh´-ber-haas-lee*). This goat was developed near Bern, Switzerland, where it is known as the Oberhasli-Brienzer, among other names.

Although some Oberhaslis were imported in 1906 and in 1920, they were not maintained as a pure breed. Not until 1936 did more purebred Oberhaslis arrive...although they were called "Swiss Alpines." They were brought in by Dr. H. O. Pence of Kansas City, Missouri.

The outstanding feature in the appearance of the Oberhasli is its rich red bay coat with black trim. The black includes stripes down the face, the ears, a dorsal stripe, and the belly and udder. The legs are also black below the knees and hocks.

Toggenburgs are the oldest registered breed of any animal in the world, with a herd book having been established in Switzerland in the 1600s. They were the first imported purebreds to arrive in this country, in 1893, and have always been popular. Poet Carl Sandburg had a well-known herd of Toggenburgs.

Toggs, as they're sometimes affectionately called, are always some shade of brown with a white or light stripe down each side of the face, white on either side of the tail on the rump, and white on the insides of the legs.

Toggenburgs rank slightly behind Saanens and Alpines in milk production, but still average over 2,000 pounds at 3.3 percent butterfat.

The **La Mancha** is a distinctly American breed. There's no mistaking a La Mancha: they look like they have no ears!

During the 1930s Eula F. Frey of Oregon crossed some short-eared goats of unknown origin with her top line of Swiss and Nubian bucks. The result was the La Mancha.

If you show La Manchas at the county fair you'll have to put up with many exclamations of "What happened to the ears!" and some people who are somewhat more knowledgeable about livestock will accuse you of allowing the animals' ears to freeze off. But you don't milk the ears, La Mancha backers say. These goats have excellent dairy temperament, and they're very productive. A good average is 1,800 pounds of milk with 3.8 percent butterfat.

Two other breeds of goats are relatively new to these shores, and at least until recently, have been considered more of a novelty than true dairy animals. However, they have become firmly established as pets, and some people do milk them.

One is the **Pygmy,** formerly called the African Pygmy. They were first

La Manchas are noted for their "lack" of ears. Many people claim this is the most docile breed, and many of them are good milk producers.

seen in the U.S. in the 1950s, and then only in zoos. These little goats are only 16-23 inches tall at the withers at maturity, and does weigh only 55 pounds. They are very cobby, compact, and well muscled, quite unlike a standard dairy animal.

In spite of this, some Pygmies are said to produce as much as 4 pounds of milk a day, and 600-700 pounds a year. And while the lactation period is shorter than for full-size goats (4-6 months rather than 10 months), the butterfat often exceeds 6 percent.

The Pygmy is more likely than the other breeds to have triplets, or even quadruplets, and the gestation period is five months.

The other newcomer is also a miniature: the **Nigerian Dwarf**. Again, while some people do milk these animals, anyone whose main interest is low-cost milk will probably start out with one of the larger, more common, and less expensive breeds. However, goat care is basically the same for all breeds, so you should find this book helpful even if you prefer the miniatures.

You might hear about a few other breeds, such as the Tennessee

Fainting Goat or Wooden Leg (which goes under several other names as well—and which at the present time has two registry associations, even though there are only a few hundred in the entire country). Angora goats have become quite popular in recent years. These are raised primarily for their long silken mohair, and while the mohair aspects are beyond the scope of a book on dairy goats, basic feeding, breeding, and management are similar.

Which breed is best?

There is no answer to that. If your reason for raising goats is to have a home milk supply, a goat that produces 1,500 pounds of milk a year is as good as any other goat that produces a like amount, regardless of breed. You might not need or want a purebred at all, at least at first. Mixed breed goats are much easier to find, they're usually cheaper, and in some cases they produce more milk than some purebreds.

Even for those interested in purebred stock, the choice of a breed isn't made because of any breed superiority or rational factors. In most cases, a breeder just "likes the looks" of a particular breed.

You'll also find it easier to find certain breeds than others, because the popularity of each varies from place to place. You might get a certain doe just because she's available, and at the same time make a wise choice because stud service will be convenient and there is more likely to be a market for kids of a locally popular breed.

This brief look at some of the basic facts about goats should help you decide if you really want to raise goats. I hope you do...but with full awareness of what will be expected of you. That means you'll want a lot more information on care and management. But before we get to that, let's take a closer look at the product that probably led you to goats in the first place: milk.

CHAPTER 2

Milk

One of the first questions a prospective goat owner who is interested in a family milk supply asks is, "How much milk does a goat give?"

While the question is logical and valid, it's something like asking how many bushels of corn an acre of land can produce. How good is the soil, how much fertilizer was applied, what variety of seed was planted, how much of a problem were weeds and insects, was there sufficient heat and moisture at the proper stages of development?

To put this in terms that might be more readily understood by city dwellers, how many ladies' coats can a merchant sell? It depends on whether the seller is in downtown New York or on the edge of a small southern village, on whether the coats are mink or cloth, whether it's June or December, and so on.

There can be no set answer to the question of how much milk a goat will give, but here are some considerations.

HOW MUCH MILK?

It must first be understood that all mammals have lactation curves that, in the natural state, match the needs of their young. Mankind has altered these somewhat through selection to meet human needs, but they're still there.

The supply of milk normally rises quite rapidly after parturition (kidding, or freshening, or giving birth) in response to the demands of the rapidly growing young. The peak is commonly reached about two months after kidding; from that peak, the lactation curve gradually slopes downward. (See Figure 1.)

This brings up what is probably the most common problem with terminology in reference to production. We often hear of a "gallon milker." The term has little or no practical value, because we want to know at what point in the lactation curve this gallon day occurred, and even more importantly, what the rest of the curve looks like. The goat that produces a gallon a day two months after kidding, then drops off drastically and dries up a short time later, will probably produce much less milk in a year than the animal whose peak day is less spectacular but who maintains a fairly high level over a long lactation. Especially in the home dairy, where a regular milk supply is the goal, a slow and steady producer is more desirable than the flashy one-day wonder.

In addition, a "gallon" is neither an accurate nor a convenient unit of measure for milk. Milk foams…and what if a goat gives just over or under a gallon? A gallon and one cup is tough to measure and tougher to record.

It's much more practical to speak of *pounds of milk per lactation*. A gallon, for all practical purposes, weighs 8 pounds. The traditional lactation period is 305 days. If a goat is to be bred once a year and dried off for two months before kidding for rest and rehabilitation, this period is logical. Even though it might be arbitrary in some cases—a goat might milk for more or less than 305 days—it's a convenient way to compare animals. Cows are judged on the same time span.

But it *is* arbitrary, and mainly for record purposes. The backyard goat dairy has no need to adhere to such a schedule, and in practice even most commercial dairies milk an animal for shorter or longer periods depending on the animal's production. In some cases it might not be worth dirtying the milk pail for a quart or so. In others, even a cup of milk might be considered valuable.

Actually, many household goat dairies with animals that exhibit long lactations would do well to milk them for two years straight without rebreeding. Production could be lower the second year, but this would be offset by avoiding a two-month layoff, breeding expenses, and unwanted kids—including the considerable amount of milk they will drink if not disposed of at birth. It should be pointed out, however, that not many goats will milk for that long: most will be dry before the ten months are out.

AVERAGES AND RECORDS

Looking at averages can be meaningless—how many American families really have 1.7 children?—but sometimes that's the only way to get even a rough idea of a situation. Just remember that when a breed averages 2,000 pounds, some of the goats are producing 3,000 pounds, and some are only milking 1,000 pounds—or less.

For many years we said a decent average was 1,500 pounds a year. There are some indications that this has been increasing. But then the question arises, is it increasing because more good goats are "on test" while less productive animals aren't? Fifteen hundred pounds—about 187 gallons—is still a reasonable expectation for the beginning goat farmer. But bear in mind that this 187 gallons in 305 days doesn't mean you can plan on 0.6 gallons a day: remember the lactation curve.

Breed records are even more meaningless for the home dairy than are averages. The new goat owner has about as much of a chance of even coming close to record production as the guitar-pickin' kid down the road has of coming up with a hit song. It takes knowledge, experience, work, and maybe even a few lucky breaks to produce a winner in any field.

At least the records will show you what goats are capable of. And they

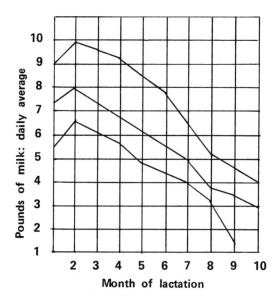

FIGURE 1.

The best doe in this herd produced 2,150 pounds of milk in ten months. The lowest record shown is 1,300 pounds in nine months. Average for the entire herd was 1,730 pounds in ten months.

FIGURE 2.

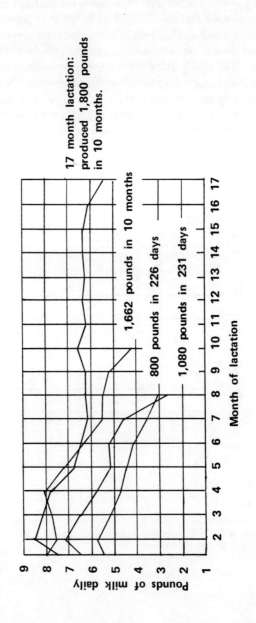

17 month lactation: produced 1,800 pounds in 10 months.

1,662 pounds in 10 months

800 pounds in 226 days

1,080 pounds in 231 days

also demonstrate what a bucket of worms you get into when you ask, "How much milk does a goat give?"

For example, the Nubian record of 4,420 pounds was set in 1972, but the runner-up produced 770 pounds less: 3,650.

The Saanen record was set in 1971 with 5,496 pounds, but in 1972 the top Saanen produced "only" 4,160 pounds and the runner-up gave 3,920 pounds.

The all-time all-breed U.S. production record was set by a Toggenburg back in 1960: 5,750 pounds. That averages more than 2 gallons a day, for 305 days. And the world record is reportedly 7,546 pounds, set in Australia in 1972. That's 3 gallons a day.

Then there are goats that freshen without enough milk to feed the barn cats.

You obviously won't start out with one of the former, and you hope you won't get stuck with one of the latter, but it would be nice to find one that's average.

Obviously these top production records don't mean a great deal to those of us who are merely interested in having fresh goat milk in our own refrigerators. It's easy to find averages, from the same breed, that vary by 40 percent or more.

The only way to know for certain how much milk a goat gives is to milk her, weigh the milk, and record it for the entire lactation period.

USING PRODUCTION RECORDS

Leaving the pacesetters for a moment, let's look at Figure 1 again. These are actual production records of a small herd of Nubians. The top doe produced 2,150 pounds in ten months, the bottom doe 1,300 in nine months. Notice the lactation curve. The average production goes from 7-1/2 pounds at kidding to about 8 pounds two months later. From there it tapers off to about 3 pounds ten months after kidding.

Figure 2 provides another example of an "average" small herd. These are actual, individual records from a herd of four grade does. It shows how much production can vary among animals. One doe had a seventeen-month lactation. She gave 1,800 pounds in the first ten months and continued to produce a steady 5-1/2 pounds daily until pregnancy caused production to drop. Another doe reached her peak in the fourth month.

If you owned these four does and were going to sell one, which one would it be? There are two lessons here:

First, remember this when you *buy* a goat: are you buying an animal someone is culling because of low production? Ask to see milk records.

Second, without records and perhaps a chart like this one, no matter how rough, you don't know for sure what's happening. Not a month from now, not a year from now, and certainly not five years from now when you're trying to decide which granddaughters of your present milking does to keep and which ones to sell or butcher.

Note that one of these does produced more than twice as much as another even though they ate about the same amount of feed and required the same amount of care. Note also that two weren't worth milking after only eight months. And finally, it should be obvious that when you milk your own goats, you don't have as steady a milk supply as when you pick up a gallon from the grocery whenever you need it. There will be times when you could drown in milk, and other times when you'll eat your corn flakes dry.

Age is a factor in milk production. Records from this same herd show that peak production comes in the fourth or fifth year.

But there are also other factors involved, and any individual goat can vary erratically from one year to the next. Even on a day-to-day basis milk production is affected by changes in weather, feed, sickness or injury, outside disturbances, and other factors.

The question "How much milk does a goat give?" can only be answered by another question: "How long is a piece of string?"

DISCOVERING A NEW TASTE

From all this you can assume that you'll be able to find a goat that will produce a respectable amount of milk for your table for at least a part of the year. Next question: Will your family drink it?

Many goat raisers delight in telling stories about how they tricked finicky people into drinking goat milk, and how those people couldn't tell the difference. In taste tests, most people will actually prefer goat milk to cow milk.

However, there have also been cases where a person has said the milk was delicious, was told it was goat milk, and suddenly decided that *really* it was awful. There isn't much hope for people like that...aside from perhaps serving goat milk from a bottle or carton that came from a cow dairy and not telling them the truth. (We're not advocating this, but some people have used this technique on their children.)

When you've been raising goats a while you'll surely be asked what you do with the milk. Many people seem to assume that goat milk is used only in hospitals. The healthful aspect of goats' milk is as legendary as their aroma and their preference for tin cans.

Many doctors do, in fact, prescribe goat milk for certain conditions. Many more would, if steady supplies of sanitary goat milk were available. It might be recommended in cases of dyspepsia, peptic ulcer, and pyloric stenosis. It is preferable to cow milk in many cases of liver dysfunction, jaundice, and biliary problems because the fat globules are smaller (2 microns versus 2-1/2 to 3-1/2 for cow milk). Goat milk has been used for infants being weaned, children with a liability to fat intolerance or acidosis, infants with eczema, pregnant women troubled by vomiting or dyspepsia, and nervous or aged people with dyspepsia and insomnia.

Goat milk is more easily digested than cow milk because the fat is finer and more easily assimilated, it is particularly rich in antibodies, and when freshly drawn it has a much lower bacterial count than cow milk. When whipped, cream from goat milk is bulkier than cow cream: the specific gravity of cow milk is 0.96 and of goat milk 0.83.

But despite all this, goat milk is not medicine: it's food. Good food! It is drunk (or used in cheese or yogurt) by more people in the world than cow milk is.

Because most Americans aren't familiar with the product, they have many misconceptions about it—misconceptions based on the comic strip image of the goat, or on the unfortunate experiences of a few who have been exposed to goat milk produced under conditions that make it unfit for human consumption.

The home winemaker who takes a basket of overripe and spoiled, wormy, moldy fruit, puts it in a dirty crock, and pays no attention to proper fermentation does not end up with a fine wine. And the goat raiser who milks a sickly, undernourished animal of questionable breeding into a dirty pail, lets the milk "cool" in the shade, and serves it in a filthy glass does not end up with fine milk. And while few people would disparage all wines after tasting the concoction just mentioned, many people are all too willing to write off all goat milk after one unfortunate experience.

Milk is as delicate a product as fine wine. It must be handled with knowledge and care, whether it comes from a cow or a goat or a camel.

Goat milk does not taste any different from cow milk. It doesn't look appreciably different. (It is somewhat whiter, because it doesn't contain the carotene that gives a yellow tinge to the fat in cow milk.) It is not richer. It certainly does not smell—if it does, something's wrong.

Most goat raisers enjoy serving products of their home dairy to skeptical friends and neighbors. The reaction is invariably, "Why, it tastes just like cow milk!"

I have noticed, however, that city people who are accustomed to regular standardized milk—milk that has butterfat removed to just barely meet minimum requirements—or worse yet, who drink skim milk, are prone to comment on the "richness" of goat milk. They'd say the same thing about real cow milk if they had the opportunity to taste it before the technologists messed around with it and turned it into chalk water. The percentage of butterfat—the source of the "richness"—varies with breed, stage of lactation, feed, and age...with goats as well as cows. On the average, there is virtually no difference between the two.

In rare cases there are animals that give off-flavored milk. We can't call it "goaty" because some cows have the same problem. It can be caused by strong flavored feeds and plants such as ragweed, grape leaves, wild onion, elderberry, honeysuckle, and many others. It can also be caused by mastitis, which often makes milk taste "salty." A rancid flavor can result when a doe is in late lactation or when foamy milk is cooled too slowly.

AVERAGE COMPOSITION OF GOAT AND COW MILK (PER 100 GRAMS)

	Goat	Cow
Water (%)	87.5	87.2
Food Energy (kcal)	67.0	66.0
Protein (g)	3.3	3.3
Fats (g)	4.0	3.7
Carbohydrates (g)	4.6	4.7
Calcium (mg)	129.0	117.0
Phosphorus (mg)	106.0	151.0
Iron (mg)	0.05	0.05
Vitamin A (IU)	185.0	138.0
Thiamin (mg)	0.04	0.03
Riboflavin (mg)	0.14	0.17
Niacin (mg)	0.30	0.08
Vitamin B-12 (mcg)	0.07	0.36

(The temperature should drop to under 45° F. in less than an hour.) A "cardboard" flavor might be caused by oxidation resulting when milk is left in the light or when there is copper or iron in the milk container. Still other flavors can come from a dirty barn and goats, and generally unsanitary practices that result in bacteria getting into the milk after it leaves the goat.

It has long been said that goat milk is "naturally homogenized" because of the smaller fat globules. Natural homogenization might be a problem for people who want cream for butter and other uses: goat milk requires a separator to extract the cream. Actually, it probably isn't the size of the fat globule that causes the cream in goat milk to remain in suspension. Research has shown that goat milk lacks a fat-agglutinating protein, a *euglobulin*. In fact, the cow is probably the only domestic animal that produces milk with this particular protein, according to Professor Robert Jenness of the University of Minnesota. Sow and buffalo milk do not form cream lines. Because homogenized cow milk is the norm in this country, most people would probably be surprised to see cream rise in milk! They won't be shocked by goat milk.

RAW MILK VERSUS PASTEURIZED

If you want to start a dandy (and sometimes heated) discussion in goat circles, just casually bring up the topic of raw milk—and then stand back.

Until recently, the idea of pasteurizing the home milk supply was seldom if ever discussed. If it was, there was usually agreement that having raw milk, and having control over its production so you could be sure of the source, was one of the *benefits* of the home dairy.

This has changed. Most health authorities come down pretty hard on raw milk, and more goat raisers are giving the question some thought. But there are still many who aren't going to start pasteurizing their milk after using it raw for years, and many newcomers prefer raw milk, too. You have to decide for yourself.

One of the potential problems is a gastrointestinal disease called campylobacter. The symptoms, which range from mild to severe, include abdominal cramps, diarrhea, and fever. Apparently farm families who drink raw milk regularly build up an immunity: most of the reported cases involved farm visitors. Raw milk advocates point out that the numbers are very small: might as well worry about being struck by a meteorite, they say.

The illness is caused by a bacteria that is universally present in birds, including domestic poultry.

It's a heated controversy, and many reasonable people see the arguments on both sides as inconclusive. You'll have to make your own decision. But if you decide to pasteurize your milk, all it involves is heating it to 165° F. for 15 seconds. Home-size pasteurizers are available. (Don't use a microwave: it doesn't work.)

CHAPTER 3

Getting Your Goat

So you've decided to buy a goat. Now the problem is finding one, and perhaps more importantly, finding the one that's right for you.

While the popularity of dairy goats has soared in recent years, they are by no means common, and they certainly aren't evenly distributed around the country. There might be many in your area, and finding the right one will be easy. Or there might not be any at all, and you'll have to do some travelling to get your home dairy started.

In 1967, there were fewer than 5,000 new registrations with the American Dairy Goat Association (ADGA), the larger of the two registries at that time. In the first edition of this book, I noted that by 1974 registrations were expected to hit 20,000—a hefty increase.

But less than ten years later, the number of goats registered with ADGA soared to 52,000!

This is just the number registered in one year, not all registered goats. It doesn't include those registered by the other associations (there are now three all-breed associations), or the vast majority of goats, which aren't registered at all.

There is no accurate count of dairy goats in the United States. While the government has recently started counting goats, many (maybe even most) aren't raised on farms covered by the farm census. According to one estimate, there are about 800,000 dairy goats in the U.S. and more than a million Angoras (95 percent of which are in Texas).

If total numbers are in proportion to the numbers of goats on official test (for milk production), it will be much easier to find a goat if you live in California: nearly 40 percent of the dairy goats in the country that are on test are in that state.

The runner-up is Oregon, but with only about 8 percent. Wisconsin is third, with about 7 percent. Other states with relatively large numbers of goats are, according to their ranking: Washington, Arizona, New York, Pennsylvania, Iowa, New Jersey, and Maine. These ten states account for an estimated 76.6 percent of the nation's dairy goat population.

WHERE TO LOOK

Don't get discouraged if you don't live in or near one of these states. There are goats from Florida to Alaska.

There are many ways to begin your search. If you have friends or neighbors who have goats, or if you have seen goats in your area, you have a good start. Almost everyone with goats has animals for sale sooner or later, and if they don't, they'll probably know of someone who does. They might even know of a goat club in your area, and then you'll have it made.

No doubt you've been watching the classified ads ever since you started thinking about getting a goat. Classified ads usually appear for only a few days, so you'll have to read them regularly and probably over a long period.

Check out the farm papers serving your area. Again, it might be a lengthy process if there are few goats in your region, but having a goat is worth it! Of course, you can also place your own "goats wanted" ad.

Be sure to attend county and state fairs, and any other goat shows. This is often a good opportunity to see several breeds at once, and also a good time and place to talk to goat people. Reach them at home and they might have a dozen other things to do, but talking about goats is one reason they're at the show! And they might have goats for sale.

There are "sale barns" in most livestock-producing rural areas. Again, the numbers of goats they handle will depend somewhat on how many goats are in the neighborhood, but even if they only see a few every couple of months, they might let you know when they have one or more if you tell them you're interested. (Be extra cautious with sale barn animals, however. Many of these are disasters, and even healthy-looking animals can pick up diseases in a sale barn environment.)

Other local sources to contact: the vo-ag teacher and FFA and 4-H leaders; veterinarians; feed stores (do they sell goat feed, and if so, who buys it?); and county extension agents.

Of course, you'll also want to watch the ads in the national goat publications. If you're lucky, there might be a knowledgeable and reliable breeder near you.

These suggestions assume you'll want to buy your first goat close to home. There are several good reasons for that. If you don't know very much about goats, you'll want to see some, and certainly the one you'll be spending a lot of time with. You'll probably want to talk to someone with experience and see their setup. You'll almost certainly want to make some close-by arrangements for breeding. If you live in a goat-deficient region, you might not have much of a choice of breeds or animals by shopping close to home, but unless you have plenty of time and money, that might be more acceptable than traipsing around the countryside.

When you gain some experience you might want a certain breed or bloodline, and will be willing to buy an animal hundreds or even thousands of miles away, maybe without even seeing it. But at first, it's much wiser and easier to do your searching closer to home.

BUYING A GOAT: SOME TERMS TO BE FAMILIAR WITH

If you're like most people who just want a family milker, you'll end up buying what's available regardless of breed, type, conformation, and even desirable traits. But you'll still want to have some idea of what to look for, and it will be helpful to at least be familiar with some of the jargon the seller might toss your way.

You're already familiar with the breeds of goats. You might encounter many other terms that help breeders identify and classify individual goats, such as registered, purebred, American, recorded grade, and grade. There is AR (advanced registry) and there are star milkers (*milkers). You might hear a seller talking about classification or linear appraisal. What do all these strange words mean, and how can you use them to help select the right goat?

A **registered purebred** animal is one with a pedigree that can be traced, through a registry association's herdbook for the breed, to the very beginnings of the recognition of the breed as "pure."

A **pedigree** is merely a paper showing the ancestry of an individual animal—a family tree. **Registration papers** are official documents showing that the animal is entered in the herd book of a registry association.

Another term you'll hear is **AR**, or **Advanced Registry.** An AR doe is one that has given a certain amount of milk in a year. The amount varies with the age of the goat and other factors, but the AR designates the animal as a good milker.

Still another term referring to production is **star milker,** or *milker. Unlike the AR, the star is based on a one-day test rather than on an entire lactation. Like the AR, points are based on the stage of lactation the doe is in. It should be noted that many does who have earned Advanced Registry certificates cannot become *milkers because they never give enough in one day. Even though points are weighted for the length of lactation, a doe still has to produce about 10-11 pounds in a day to earn a star.

Conversely, many *milkers can't earn their AR because they don't produce enough in the entire 305-day lactation to qualify. Both are official, however, and both are supervised by someone other than the owner.

The star is based on total pounds of milk produced in 24 hours; the total number of days in the current lactation, based on 0.1 percent for each ten days with a 3 point maximum; and the butterfat percentage. A doe that gives 6.4 pounds in the morning and 7.0 pounds in the evening would earn 13.4 points. If the butterfat is 3.7 percent, 3.7 times 13.4 equals 0.4958 pounds of fat. Butterfat is divided by 0.05 for point purposes: in this case she gets 9.91 points. If the doe was fresh 44 days, this adds 0.44 points for lactation. The total would be 23.75 points. A minimum of 18 are needed to earn the star.

If a doe's dam has earned a star and she earns one herself, she becomes a **milker. If her granddam also had a star, she is a ***milker.

Bucks can also have stars, signifying those earned by their maternal ancestors.

Some purebreds are not registered, for a variety of reasons. The breeder might not think it's good enough to warrant registering. In many cases, smaller breeders who aren't really interested in registered animals simply don't want to bother with the paperwork and expense. In some cases, with a lot of work or luck or both, you can trace a purebred but unregistered animal back to its registered ancestors, and with the proper paperwork, have it registered. More commonly, it's impossible to prove that the goat truly is a purebred if it isn't already registered.

An animal without a pedigree is considered a **grade.** It could be purebred, but without the papers, you can't prove it. Most grades are mixtures of two or more breeds.

Some grades are very good animals. If such a goat meets certain requirements, it might be listed as a **recorded grade** of whichever breed it most closely resembles.

If such a recorded grade doe is bred to a registered buck, the offspring will be 1/2 "pure." The does from these matings can be recorded grades. If one of these does is bred to a purebred buck, the kids will be 3/4 purebred. One more generation will result in a goat that is 7/8ths pure, and one more will produce kids that are 15/16ths pure.

These can be registered as **Americans** (7/8ths for does, 15/16ths for bucks). You might also encounter **NOA** (Native on Appearance) **Americans** and **Experimentals** (which usually means a buck jumped the fence and bred a doe of a different breed). And of course, in some parts of the country people will talk about brush goats, scrub goats, Spanish goats, fainting goats, and even just plain old nanny goats.

CHOOSING THE RIGHT GOAT FOR YOUR NEEDS

Assuming you have a choice of goats from all these classifications, which is best?

Once again, there is no simple answer. Here's why.

If you're keen on showing, you'll want registered purebreds. But our emphasis here is on milk. For a home dairy, a registered goat, or even an unregistered purebred, may or may not be the best choice.

Registration papers don't mean anything more than that the goat is listed with one of the registry associations and that the pedigree can be traced back to the closing of the herd book. This is important to experienced breeders who are trying to upgrade and improve their animals. But if all you want is milk, and you know nothing about goats' family trees, the papers don't mean much. In fact, some registered purebreds are very poor milkers. A registration certificate is not a license to milk.

On the other hand, many grades, and even brush goats or scrubs, turn out to be excellent milkers. One illustration of this comes from a lady in a southern state who bought a brush goat. It appeared to be of good Toggenburg breeding, but it had been running semi-wild, as brush goats do, clearing a patch of hilly land so cattle could graze there. These goats are not fed hay or grain, and they aren't milked.

When this particular doe was treated as a *dairy* goat, she turned out to be a superb milker! She had always had the genetic ability, but had fallen into the hands of someone who didn't know or care. With proper feed and management, she blossomed. There are many "common" goats like this.

In other words, you might find very fancy, nice-looking, papered goats that won't produce enough milk to feed the cat...and you might find rather ordinary, crossbred, inexpensive goats that will fill your gallon milk pail to the brim. Your problem as a prospective buyer is, how do you know if a particular goat will be a reliable and efficient milker?

This is a two-prong question. We want to know if a goat will produce milk, and we want to know if she will be efficient, which is to say, economical.

Registration papers and pedigrees might tell you something about her milking ability, but only if you know how to read them. Show wins (of the animal herself or of her ancestors) may or may not means she's a good milker. You might look at stars, AR certificates, classification scores, linear appraisal scores, DHIA records, and more... but a goat that has none of these isn't necessarily a poor milker. It might just mean that the owner hasn't bothered to go after them.

The upshot is that all of these records and papers give you some degree of insurance—not a guarantee, but some insurance.

But all of these take time and money. You pay for the insurance. Only you can decide if it's worth it, for any particular animal. If you pay hundreds, even thousands of dollars for a good goat, your milk will be that much more expensive.

It's even more complex than that. As we'll see in Chapter 14, you might *lower* your milk costs by having registered purebred kids to sell, if you can sell them for a good price. On the other hand, as a beginner, you might be more comfortable "learning" with a less expensive and perhaps more adaptable animal. To all this add the fact that sometimes there are very good purebreds for sale at very reasonable prices, and rather poor grades for sale at exorbitant prices!

CONSIDER THE SOURCE

How do you cut through this maze of confusion and conflict? Very much like you buy a car or anything else. You arm yourself with as many facts as possible; you rely on the integrity and reputation (and probably even the personality) of the seller and the seller's place of business; and then you buy a particular model just because you like the color or because it "appeals" to you!

First, it helps to find a breeder you can have faith and confidence in. You're going to have to rely on that person's experience and honesty and integrity, to a large degree, for your first purchase.

Are the goats running loose or staked out in a weed patch, or are they well housed in a neat and comfortable building and an exercise yard with good fencing?

Are the animals disbudded, neatly trimmed, and with hooves that show obvious care?

Does the owner appear to be knowledgeable about goats? Trustworthy?

If there are papers or other official records, fine. If there are only "barn records," kept by the owner but without any official standing, you'll have to rely on the owner's honesty. If there are no records at all, it could mean that the seller doesn't know very much about goats or doesn't care very much about them.

In any case, it will be helpful to have some idea of the differences between a "good" goat and a "poor" one. This will take a great deal of experience, but here are some general tips on getting started.

CONFORMATION

The general appearance of livestock—the way an animal is put together—is called "conformation." Conformation is what a dairy goat judge is looking at when placing animals in the show ring. While a licensed judge spends many hours and years of study and practice, to a certain degree you must be a "judge" when you buy a goat and as you build and improve your herd.

This means you must know the parts of the animal. You must know what good animals look like and what traits are considered defects. You must be observant enough to see both good conformation and defects, and knowledgeable enough to weigh and evaluate their relative importance. Some people have a sixth sense for evaluating livestock: others never get the hang of it. While you'll have to see and handle hundreds of animals and study far beyond the scope of this book to become even a fair judge of goats, here are some ideas you should be familiar with even when you purchase your first doe.

First observe the animal from a distance. Note her **general shape** and carriage. She should be feminine, with a harmonious blending of parts. The scorecard speaks of "impressive style, attractive carriage, and graceful walk," but this is no beauty contest! These traits can tell a great deal about her general condition, vigor, and the dairy character that means milk in the pail.

Then move in for a closer look.

The **head** should be moderately long with a concave or straight bridge

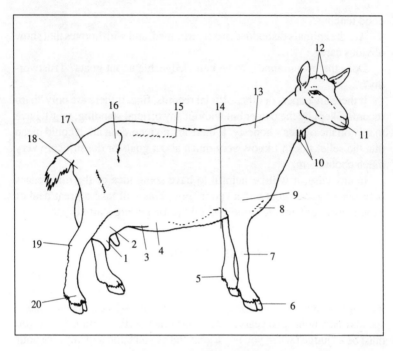

The Parts of a Goat's Body

1. teats	6. hoof	11. muzzle	16. hip bone
2. udder	7. knee	12. knobs	17. rump
3. milk vein	8. chest	13. withers	18. pin bone
4. belly	9. shoulder	14. heart girth	19. hock
5. claw	10. wattles	15. back	20. pastern

to the **nose,** except in the Nubian, which must have a definite Roman nose. Saanens have a concave nose, or dish face. The **eyes** should be bright, the forehead between the eyes broad.

Ears are a part of conformation, but they're of small importance if your primary concern is having fresh milk. But to make this description more complete, note that the ears should be pointing forward and carried above the horizontal, again with the exception of the Nubian, which must have a long, thin-skinned ear, hanging down and lying flat to the head. La Manchas, the "earless" breed, have a size limit of about an inch (2.5 cm) for the ear. The so-called airplane ears that result from a cross between a Nubian and another breed are ridiculed by many, but again, you don't milk the ears, and more than a few people have such animals and love them.

Of more importance to the home milk supply are such points of conformation as the **muzzle,** which must be broad with muscular lips and strong jaws, as this is an indication of feeding ability. Large, well-distended nostrils are essential for proper breathing.

The **neck** should be clean-cut and feminine in the doe, masculine in the buck, with a length appropriate to the size of the animal. It must blend into the shoulders smoothly and join at the withers with no "ewe neck." The goat needs a large, well-developed windpipe.

The **forelegs** must be set squarely to support the body and well apart to give room to the chest.

The **rib cage** should be well sprung out from the spine with wide spacing between each rib. The chest should be broad and deep, indicating a strong respiratory system. The back should not drop behind the shoulders, but should be nearly straight, with just a slight rise in front of the hip bones.

The **hip bones** should be slightly higher than the shoulder. The distance between the hip bones and the pin bones should be great, but not so long as to make the animal look out of proportion. The slope of the rump should be slight and the rump should be broad. The broader the rump, the stronger the likelihood that the goat will have a high, well-attached udder.

The **barrel** should be large in depth, length, and breadth. A large barrel indicates a large, well-developed rumen necessary for top production.

There are many types of **udders and teats.** Abnormalities such as double teats, spur teats, or teats with double orifices are to be avoided. (On very young kids, extra teats can be clipped off.)

While very large "sausage teats" are undesirable, very small ones may be worse, as they make milking difficult, especially for people with large hands. However, many first fresheners have tiny teats that quickly become more "normal" with milking. It's often easiest to let the kids nurse does like these.

Don't be too impressed by large udders. Many of them are just meat. With a very pendulous udder you'll have to milk into a pie pan because there isn't room to get a pail under the goat! Of more serious concern, pendulous udders are more prone to injury and mastitis infections.

A well-attached, capacious udder, carried high out of harm's way, with average size teats and free from lumps and other deformities, is the heart of your home dairy.

The condition of the **skin** reflects the general condition of the entire animal. It should be thin and soft, and loose over the barrel and around the ribs. A goat with "unhealthy-looking" skin probably really isn't very healthy, or it might have internal or external parasites. Check for lice and mites.

If you're looking at younger stock, **avoid the overdeveloped kid.** A kid that develops too early seldom ends up being as good an animal as one that has long clean lines and enough curves to indicate that the framework will be filled out at the appropriate time.

Many goats have **wattles,** which are small appendages of skin usually found on the neck, although they can be just about anywhere on the body. They are a family trait, not a breed characteristic: some animals of all breeds have them, others don't. They are merely ornaments. Some breeders cut them off young kids not only to make the animal look smoother but because sometimes another kid will suck on them and cause them to become sore. Cut with a sharp scissors at an early age, they will bleed but little, or you can tie a thread tightly around the base and they will fall off.

Horns, likewise, are indicative of neither sex nor breed. Some goats have them, some don't. And many goats born with horns have them removed soon after birth because horns can cause many problems later. Disbudding young kids is much easier and safer than dehorning older goats. See Chapter 7 for more details. These ornaments are a disqualification for dairy goats that are shown.

Then observe the goat from the side, the front, and the rear with a critical eye. A good dairy animal has a classic **"wedge" shape** when viewed from above: she has a delicate neck rather than a bull neck, and the barrel is wider than the shoulders. The top line is straight, and a severely sloping rump is a defect.

In official judging, general appearance and breed characteristics are allotted 30 points; dairy character, 20; body capacity, 20; and the mammary system, 30; for a total of 100 possible points.

CHECK THE RECORDS

If for you a "good" goat is simply one that milks well, first consider health, conformation, and overall appearance, because a sickly animal or one that isn't built like a dairy animal isn't going to do the job for you.

Next consider any records that might be available. Most people think official milking records such as DHIA (Dairy Herd Improvement Association) records are best. DHIA was set up as a management tool for dairy cows: its adoption by goat owners is relatively recent. It's administered by the USDA through state extension services. Basically it consists of monthly surprise farm visits by a trained tester who weighs the milk and determines the butterfat content. The records are unbiased and official: you don't have to take the owner's word for how much a goat produces.

A variation known as *group testing* is common among goat owners, because it's cheaper. In group testing, an official, paid, DHIA tester makes one surprise visit a year. The monthly testing is done by the group members themselves, on a rotating basis: I test your goats, you test mine.

DHIR (Dairy Herd Improvement Registry) is the same as DHIA, but these tests also apply to ADGA records.

Even the owner's own barn records of daily milk production are better than none, however. Remember that the amount of milk produced in a year is more important than the amount produced on any one day. You don't expect your first goat to break any world records, but you want her to be more than a nonmilking pet, too!

Show wins can be impressive, and they'll tell you something about what qualified dairy goat judges think of the animal's conformation, but like registration papers, blue ribbons are no license to fill a milk pail. Registration papers and pedigrees mean little unless you're familiar with a great many names and backgrounds, and that won't come until later, after study and experience.

The easiest way for the beginner to judge all of these papers and records, including barn records, is to judge the character of the seller. It might be more difficult to judge the owner than the goat, but at least you have a lifetime of experience with people! Without records of any kind, your best assurance of getting the goat that's right for you is by buying from someone you feel you can trust. Such a person will help you learn, and some will even stand behind the animals they sell (although this is sometimes asking a lot: for instance, when careless or ignorant people take home a good goat, neglect or abuse it, and then complain that they were ripped off because the animal doesn't meet their expectations). This is the kind of seller you can call on later for help and advice. When you deal with someone like this, you get much more than just a goat for your money.

On the other hand, there are people who have been raising goats for years and who still don't know as much as you will after reading this book. While some people have twenty years' experience, these folks have one year's experience twenty times. There are also out-and-out crooks, who will sell worthless animals and overpriced animals. Some will sell you "registered" goats with the papers to come later (they never do), and there are others who are more interested in disposing of a goat or acquiring your cash than they are in helping you or promoting goats.

People who raise goats, in other words, are a cross-section of people in general. If you buy a goat without knowing very much about goats, it will help to know something about people.

There's a reverse side to this, too. Some *buyers* are pushy, obnoxious, know-it-alls, stupid, and crooked! Not you, of course, but bearing in mind that an experienced seller has most likely dealt with people like this in the past, be aware that *you* are being evaluated, too, and act accordingly.

OFFICIAL RECORDS AND BARN RECORDS

Official milking records will be in pounds and tenths, and unofficial barn records should be, too. For all practical purposes a quart of milk weighs 2 pounds; a gallon is roughly 8 pounds. Be wary of milk records expressed in pints and quarts (and downright skeptical of milk recorded in gallons!). Even an honest and well-meaning milker can be misled by a bucket of foaming milk that "looks" like 3-1/2 pints. Weight is much more reliable.

Barn records depend entirely on the accuracy of the scales and the integrity of the milker. They can be falsified or altered. Official tests, monitored by outsiders, are much more reliable. Because of cost and other factors, they aren't widely used by goat owners, although they're becoming more common in some areas.

With or without papers—registration certificates, pedigrees, show wins, barn or official test records, advanced registry certificates or stars—many people recommend that prospective goat buyers see the goat being milked, or better yet, milk her themselves. For an inexperienced milker, the doe will probably be nervous and you might not get much milk. But it's better to get a lesson, even if the owner has to finish the job, than to get home and find out you can't milk. Taste the milk to check it for off flavors.

ASSESSING THE GOAT'S WORTH

You've found your goat, and you're ready to deal. Next question: How much is a goat worth?

Once again, there are no set answers. The price of goats generates as much heated conversation among goat people as anything else. Some plug for higher prices, some for lower prices, and there are good arguments on both sides. Goats have been sold for more than $3,000, and of course, many more have been *given* away.

For the person whose primary interest in goats is an economical milk supply, there's a way to determine what an animal is worth. But it will take some work, and it isn't foolproof. It involves guesstimating how much your home-produced milk will cost.

Even a formula approach isn't really much help, however. It would be impossible to fill in the blanks of a formula in a book, because hay and grain prices vary widely from one area of the country to another, and so do milk prices. They also vary from year to year: in some places they can double after a drought or crop failure. But here's an example anyway.

My cost for a prepared grain ration right now is $16 per hundred pounds, or 16¢ a pound. Hay is being quoted at $165 a ton, or 8.25¢ a pound. If I feed my goat approximately 3 pounds of grain and roughly 3 pounds of hay a day over the course of a year, I'll need about half a ton of each. The grain will cost me, at current local prices, $160, and the hay will cost in the neighborhood of $85, for a total feed cost of $245 a year.

If this goat meets my expectations and produces 1,500 pounds of milk, my feed costs for the milk will be just over 16¢ a pound, or about $1.30 a gallon.

Goat milk in a health food store would cost me, in this time and place, a little more than $6 a gallon, so the 180 or so gallons from my goat would have a retail value of more than $1,100! Ignoring incidental expenses, and certainly labor, I could pay more than $800 for the goat and get my money back within a year. The next year that could be considered clear profit, and if she has two kids....

But wait a minute. Much as I prefer goat milk, what if $6 a gallon is too much to pay, so I opt for cow milk, which currently sells for $2.19 a gallon? Although we're comparing apples and oranges now, my goat milk would only have a value of roughly $400. I could figure on making $155 above feed costs.

But I might not really use that much milk, or my usage requirements might not fit the goat's lactation curve. There are many personal factors to consider. And of course, never forget that goats are living creatures, not machines: no one can predict if or when they'll get sick, or even die, or how much milk they'll produce.

If economical milk is your real concern, or if you enjoy playing with the spreadsheet program on your home computer, you can spend a lot of time juggling the numbers. By taking into account the life expectancy of the goat and the value of her kids during that period, as well as breeding, veterinary, and other expenses, you can get a fair idea of what a goat will be worth to you. You can also determine how much more you can afford to pay for a goat that produces 500 or 1,000 pounds more—or less than—1,500 pounds a year. You'll want to come back to this when you start to upgrade, or when you wonder what effect reducing feed costs would have on your milk bill.

The real purpose of bringing all this up at this particular point is to demonstrate that anyone who thinks a decent dairy goat isn't worth more than $25 or $50 isn't being very realistic. It costs more—in some places much more—than $100 just to raise a kid to milking age. If you find a seller who doesn't know that, or doesn't care, and will sell you a nice goat for little money, fine.

But if a responsible breeder asks for a price in line with what the goat is really worth, don't go into shock.

And if a seller wants to charge what you think is a ridiculous sum, go back over this chapter again and plug in the details relating to your own specific situation.

In the final analysis, chances are you'll get a goat just because you want a goat! And just like with a new car, you'll end up bringing home the one that fits not only your budget, but your personal tastes—and your own unique personality.

———————————— CHAPTER 4 ————————————

Housing

Many people, especially those who haven't had much experience with livestock, are prone to bring home an animal and *then* decide where and how they're going to keep it. This is definitely starting off on the wrong foot.

Most people contemplating raising goats already have facilities that, with a little work, will serve as a shelter. (Anybody new to goats would be well advised to learn something about them—not just from books, but from practical experience—before building brand-new facilities. A few years' experience will go far toward eliminating costly mistakes.)

Goats do not have to be kept warm even in northern climates if they've been conditioned to the cold through the fall. But in any climate, their housing must be dry and free from drafts. Goats are very susceptible to pneumonia.

Goats are commonly kept in garages and sheds, old chicken coops and barns. These may be wood, concrete, cement block, or stone. The floors might be wood, concrete, dirt, sand, or gravel. Although goat owners like to argue about the relative merits of each, goats are doing just fine in all of them.

Ideally, the goat house should be light and airy with a southern exposure. It should be convenient to work in, which means the aisles and doorways should be wide enough to get a wheelbarrow through without barking your knuckles; feed and bedding storage should be conveniently nearby; and running water and electricity should be available to make your work easier, more pleasant, and safer. Seems like most of the features of the "ideal" goat house are more for the benefit of the goat farmer than for the welfare and comfort of the goats!

FLOORS

WOOD

Wooden floors such as those found in brooder houses or other poultry buildings can be warm and dry if the rest of the structure is snug and tight. But wood rots. This means highly absorbent bedding should be used, and it should be changed frequently. The most absorbent bedding is peat moss, which can absorb 1,000 pounds of water for each 100 pounds of its dry weight, far more than any other bedding material. But if you have to buy it, it's expensive, which means it isn't very widely used.

Chopped oat straw rates second among commonly used materials, but it only absorbs 375 pounds of water per 100 pounds of dry weight. Note that this is *chopped* straw: long straw will only absorb 280 pounds per hundredweight. Wheat straw is somewhat less absorbent than oat straw.

Wooden floors are obviously not the most desirable for animals like goats, where large quantities of wet bedding will accumulate. You wouldn't put wooden floors in a new building designed for goats. But if you already have the building, there's no reason why you couldn't use it.

CONCRETE

Concrete floors are only somewhat less desirable, according to many experienced goat raisers. Concrete is cold. The urine cannot run off—which of course is just fine with many people who prize the used bedding for their fields and gardens or compost bins. Concrete floors do require a great deal of bedding, but I don't consider this a serious drawback unless bedding is very expensive and/or you don't have a garden to use it on.

Cow manure is extremely loose and liquid, certainly in comparison with nanny berries. Those neat little compact balls bounce, and in this context, bouncing is preferred over splashing. With a little fresh bedding to keep the top surface clean, goat litter can accumulate to a considerable depth and still be much less offensive than a cow stall.

The problem with concrete floors is that while the surface might be quite clean and dry, the bottom layers can be a swampy morass. Deep litter does not signify a sloppy goat farmer: for goats, the deeper the better, certainly on concrete floors, and especially in winter. The lower layers will actually compost in the barn if they aren't too wet, not only helping to warm the goats' beds in the same way the old-fashioned hotbeds warmed early garden seeds, but also hastening its use in the garden. Some people even use compost activator on the bedding to speed up the bacterial action.

Such deep litter is warm, and quite odorless (until you clean the barn, that is).

Concrete floors are easy to get really clean, which can be very important in the summer when deep litter might not be so desirable. Goats prefer less bedding in the warmer months, but even then concrete floors require special management. Concrete is always relatively cool, and hard, and there have been reports of rheumatism-like problems arising among goats on hard surfaces. Sleeping benches—simple raised wooden platforms—can be used both winter and summer, and the goats are enthusiastic about these regardless of floor type.

I have always had concrete floors, and I like them. But other people have actually torn out concrete to install what they consider the ideal: dirt or gravel floors.

EARTHEN

Earthen floors are the easiest to maintain. The urine soaks away, and much less bedding is needed. This is an important consideration for people for whom bedding is expensive and who don't have gardens to use it on. Of course, if you hate to see those nutrients leach away, and you want compost or mulch for your garden anyway, you'll have a different view.

Soil is warmer than concrete and more comfortable for the animals, at least if only a small amount of bedding is used.

Your preference depends upon your situation. Some years ago there was a lady who kept goats in what was practically the center of town, as the suburbs grew up around her. She kept the animals on concrete floors, without bedding. The urine drained away, and the droppings were swept up daily. The place was spotless, there was never a complaint from the neighbors, and the only waste disposal was a daily coffee can of nanny berries that went on the rose bushes.

The floor and its proper maintenance will contribute a great deal to the health and comfort of your goats.

INSULATION

We'll assume that the roof doesn't leak and that drafts aren't getting into the house through cracks in the walls and around windows and doors. We do want ventilation, but not *drafts*.

Insulation can be highly desirable, but special precautions will have to be taken to protect it from the goats. Most wall materials will be chewed, eaten, or smashed. Plywood, plasterboard, and the like won't last more than a couple of days. Stout planks are fine, and cement wallboard can also be used. Naturally, you wouldn't use plastic sheeting or similar materials the goats could chew on with dangerous consequences.

PENS

Most goats today are raised in what is called "loose housing." That is, instead of being chained in stalls or kept in individual box stalls or stanchioned like miniature cows, they are free to move about in a common pen. While many cow dairies have been converting to this system to save labor, it makes even more sense for goats.

Goats are herd animals. They need companionship. They are active animals and need exercise. And loose housing is a whale of a lot less work than individual box stalls or tie stalls. Loose housing obviously entails lower original cost in both construction time and materials, and it's more flexible: if you have four stanchions, there's no way you can house five goats.

SIZE REQUIREMENTS

The size of the building is dependent on several factors. Recommendations range from 12 to 20 square feet per animal (the lower figure in warm climates where they spend more time outside). If there is a sizeable pasture or exercise yard and the barn is used mainly as a dormitory, you could get by with a smaller size. But adequate space for the animals themselves is only one consideration.

You'll need space to store hay, bedding, and grain, and more likely than not the milking will be done in the barn, too. If you're working with minimum space requirements, don't forget that the addition of kids will require more room. And then, too, there is always the possibility that your herd will increase. Goat herds have a way of doing that, even on well-managed and very small places!

In other words, if you grow your own hay and straw or buy it off the baler when it costs less and will store a year's supply, you'll naturally need more room than if you intend to bring home a week's supply at a time. If you spend more money to have a larger building, you can hope to save some by buying feed and bedding in large quantities.

If you stock up on hay and straw during the harvest, roughly half of this will be gone by kidding season. Consider planning a combination hay storage and kid raising area.

Here again, if you already have facilities that aren't perfect, don't panic. If there simply isn't room for hay in the building you'd like to use for goats, it can be stored elsewhere. It will involve more labor, but it won't keep you from raising goats. Grain can be stored in garbage cans in outbuildings, if necessary. Milking elsewhere can be extremely inconvenient, however, especially in inclement weather. Unless an alternate shelter is very close to the goat shed, try to incorporate milking space into the main building.

Grade A dairies, cow or goat, *must* have a separate room for handling milk. Where large numbers of animals are involved, a separate milking parlor is advisable to eliminate dust and barn odors. But for the backyarder or homesteader with a few animals that are kept clean, milking in an aisle is acceptable, and far more convenient.

A milking bench isn't a necessity, but it will contribute greatly to the ease of milking and will result in a cleaner product.

The major item of equipment in a goat house is the manger. These can be constructed in many ways.

Goats are notorious wasters of hay, and this is the main factor to consider when designing and building a manger. Grain is often fed in mangers, since most goats won't get their allotted ration during milking. Greed can cause problems, then, unless some means of fastening the animals into the manger is devised to prevent bossy does from taking more than their share. Hay can be fed free choice in the same manger: the goats can come and go as they please and eat as much as they want.

GATES AND LATCHES

There is much variation in jumping ability, or perhaps desire. Contented goats are less likely to leap fences of any height. But if a deep litter system is used, remember that a 4-foot fence in October might only be 3 feet high a short time later! (And naturally, as the floor goes up, the ceiling comes down, an important consideration if it's already low or if you're tall.)

Gates and latches are important in goat houses. Gates should be sturdy, for goats love to stand on things with their front feet and gates are the favored place to do this standing. With deep litter, the gate should swing out of the pen, which is a good idea in most cases anyway. Make sure the gate is wide enough to get through with a wheelbarrow or whatever you'll

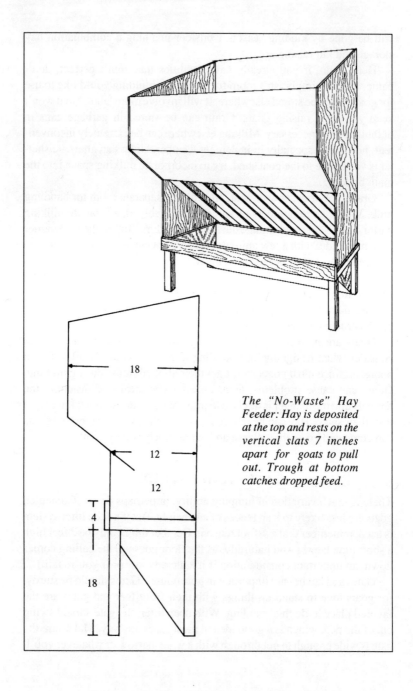

The "No-Waste" Hay Feeder: Hay is deposited at the top and rests on the vertical slats 7 inches apart for goats to pull out. Trough at bottom catches dropped feed.

be using at cleaning time. Since most goats are Houdinis when it comes to unlocking latches, pay special attention to those.

UTILITIES

While some people, goat raisers or not, demand more luxuries and conveniences than others, two are of particular interest here: water and electricity. Water piped to the barn can save countless minutes, which on an annual basis amounts to hours or even days. The goats are more likely to have a continuous supply of fresh water if you don't have to lug it a long

A manger is necessary for feeding hay and grain. Goats are allowed to eat as much hay as they want, but the grain is rationed: usually 1 pound per day for maintenance and 1 additional pound for each 2 pounds of milk produced.

18 X 18 GOAT BARN

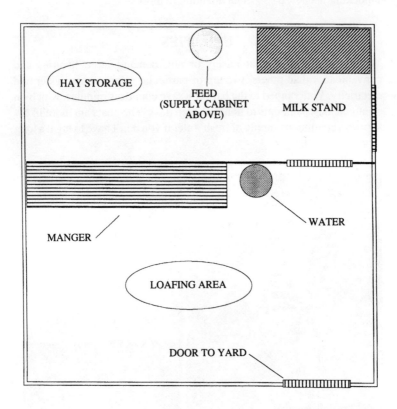

The simpler the barn plan, the better in terms of original cost and ease of cleaning and maintenance, as well as convenience, when doing daily chores. Straight lines cost less to build, and since corners are hard to clean, try to keep them to a minimum.

A plan such as this could comfortably house four or five does. Provisions could be made for kids in the hay storage area, or in a temporary pen within the loafing area.

Of course, the dimensions can be changed to accomodate more goats, fewer goats, and more or less hay.

If space and construction funds are limited, a milk bench that folds up against the wall will save space; hay can be stored elsewhere or brought in on an as-needed basis; the manger and water could be placed under a simple roof outside.

There's plenty of room for imagination when constructing goat facilities.

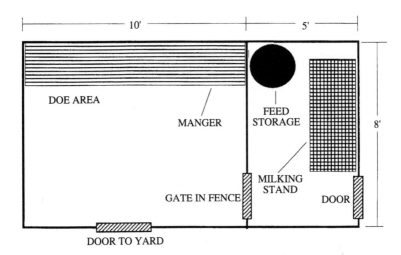

Barn layouts for goats can be very complex—or very simple. Simple is better, especially for beginners: get some personal experience before you spend a lot of money on a building. Something as simple as the above is all you or your goats need, and many people even make do without the milking area and stand.

distance, and from that standpoint alone the plumbing can be worthwhile. A hose might work in summer or in a warm climate, although it's an unsightly nuisance, but where freezing occurs, buried pipe and a frost-proof hydrant can almost be considered necessities.

Electricity is obviously a boon when you have to do chores before or after the sun shines. Trying to milk or deliver kids by flashlight is hectic, and lanterns can be dangerous as well as a bother. Moreover, you'll eventually want electricity for clippers, disbudding irons, heat lamps, and perhaps for a stock tank heater to keep drinking water from freezing, as well as other possibilities.

OTHER CONSIDERATIONS

You'll want storage space: how much depends on the type of operation you have. There should at least be room for a pitchfork out of harm's way, hair clippers, hoof-trimming tools, brushes, disbudding iron or caustic, and perhaps a medicine cabinet. Provide a place for a hanging scale, and make sure the milk records are kept where the goats can't nibble on them! (This is the voice of experience speaking.)

Buildings should be whitewashed or painted with a lead-free white paint. This will make the building not only more attractive and pleasant to work in, for you and the goats, but it will be cleaner, and light colors tend to discourage flies and other pests.

Milking equipment will be stored somewhere cleaner than the barn, of course. Most home dairies use the kitchen as a milkhouse. The ideal milkhouse—well ventilated, with hot and cold running water, rinse sinks, floor drain, impervious walls and ceiling—is nice, but a bit much to expect for a dairy with only a few goats. The kitchen works just fine for most people. That's where the utensils and strainer pads are kept, the milk strained and cooled, and all milking utensils washed.

This chapter includes some suggested floor plans. They're only suggestions: you'll want to adapt them to your own particular circumstances. Don't get too fancy or spend too much money right away if you don't have any experience with goats. After a few months of doing chores, you'll probably want to make some changes based on your building, your animals, and the way you do things.

Facilities for dairy goats need not be elaborate or expensive. But because they are dairy animals, you'll want to keep them and their surroundings as clean as possible; and because nobody does unnecessary work just for the fun of it, and because you want the best production your does are capable of, you'll want to plan quarters that are easy to keep clean, that are pleasant for both you and the goats to be in, and will contribute to the health and well-being of your herd.

CHAPTER 5

Fencing

Fencing is probably more important—and more difficult—with goats than with any other domestic animal. Goats will jump over, crawl under, squeeze through, stand on and/or lean against, and circumvent any boundary that is not strictly goat-proof in any other way they can discover or invent.

For most people a lot of fencing really isn't necessary. In most cases, you won't want to think in terms of "pasturing" your goats. Even if you have plenty of land, goats won't make good use of the usual pasture plants, grasses and clovers. Goats prefer browse: trees and shrubs and brush. Goats that are fed in the barn will probably ignore even the finest pasture...although they'd be delighted to get at your prize roses, specimen evergreens, and fruit trees!

In most cases a small, dry, sunny exercise yard is sufficient.

For many people the picturesque board or rail fence comes to mind first. It won't work for goats unless it's all but solid, because they can slip through openings you wouldn't believe.

Woven wire is less expensive, but that has drawbacks, too. If your goats have horns, they'll put their heads through the fence, then be unable to get free. Worse, they'll stand on the wire, or lean against it, until it sags to the ground and they can nonchalantly walk over it. Even with close spacing of posts and proper stretching (the latter a crucial part of building this type of fence), woven wire will soon sag from the weight of goats standing against it and will look unsightly, and eventually be useless.

Avoid barbed wire. It's awfully mean stuff around tender-skinned, well-uddered goats. And it doesn't impress them anyway.

Avoid picket-style fences where a goat can stand against the fence

Right: *Premier Fence Systems's seven-wire, electrified high-tensile fence.*

Left: *This ten-wire, nonelectrified wire fence can also be electrified.*

with her front feet, slip, and impale her neck on or between pickets.

The ideal goat fence for the small place would probably be **chainlink,** but like most ideals, it just isn't in the budget for most of us. A very good and somewhat less expensive alternative is "stock fencing," which is made of welded steel rods and comes in a variety of lengths and heights. For goats it should be 4 feet high, and sections 16 feet long are easy to handle because the stuff is fairly lightweight. Stock fencing can be attached to regular steel or wooden fence posts.

Electric fencing should be used much more than it is for goats. They have to be trained to respect it, but once they know what happens when they touch it, it's possible to fence even large areas at low cost. Train them in a small area. Until they get zapped once or twice, they'll be crawling under, jumping over, and just plain busting right through.

Field fencing (woven wire) with a strand of electrified wire just inside it at about nose height makes a very good goat barrier. The electric fence keeps the goats from reaching the field fencing, and the field fencing offers more security than the hot wire alone.

A recent addition to the possibilities is what is often called **New Zealand-type** fencing. This is powered by a 12-volt battery and an energizer that, when properly grounded, intermittently sends out 5,000 volts. It uses fiberglass posts, except on corners.

Good fencing is a necessity for goats, and fencing any sizeable area entails a large investment. In most areas, the nourishment the goats get from a pasture won't justify the cost of the fencing. Build a small exercise area—and build it well—and bring the feed to the goats.

CHAPTER 6

Feeding

No aspect of goat raising is more important than feeding. You can start out with the very finest stock, housed in the most modern and sanitary building, but without proper feeding your animals will be worthless.

The proper feeding of goats requires special emphasis for several reasons. High on this list is the fact that many people who start to raise goats have little or no experience with farm animals. Feeding goats is a lot different from feeding cats or dogs or parakeets. Goats are ruminants, which affects their dietary needs, and unlike the average cat or dog, they are productive animals, which puts additional strain on their bodies and requires additional nutriment. And with cats and dogs or other household pets, there is no temptation to toss them an armload of grass clippings or cornstalks from the garden!

COMMERCIAL FEEDS OR MIX YOUR OWN?

A discussion of feeds can be very long, or very short. It can be short if you buy a commercially prepared grain ration and follow the directions on the label. It will have to be long if you mix your own because that will require at least an awareness of the basics of nutrition, physiology, bacteriology, math, and more.

There is no middle ground. You can't, for example, arm yourself with a "goat food recipe" and do a good job of feeding. For one thing, the feed value of hay and grain varies from place to place and year to year, being affected by soil, climate, and other factors. Pacific Coast-grown grains are lower in protein than those grown elsewhere; hay harvested at the proper stage of development and well cured will differ dramatically from hay that is cut too late and leached or spoiled by improper curing, so far as nutrients are concerned. (Moldy hay should never be fed.)

Just as important, any given ration depends on locally available ingredients and their comparative prices, and the suggested rations almost invariably have to be adjusted. Unless the feeder knows what to look for, the carefully formulated suggestions will be thrown out of balance by indiscriminate substitutions.

Likewise, even the person who feeds commercial rations can destroy the balance by haphazardly adding "treats" or by making use of available grains in addition to the commerical feed. You can no more prepare a balanced diet by adding a handful of this to a scoop of that than you could expect to bake a cake by using the same method.

Many people, shocked by the cost of commercially prepared bagged feeds, develop an interest in mixing their own. After all, they reason, if you can buy 100 pounds of corn for 5 or 6 dollars, why pay $15 or more for 100 pounds of bagged feed?

In recent years there has been a great deal of interest in "organic" feeds—feeds grown without chemical fertilizers, herbicides, or other pesticides, and processed without antibiotics, preservatives, medications, and so on.

Then there are people who grow hay and grain anyway. Naturally, they'd rather feed that to the goats than buy a commercial mix.

THE DIGESTIVE SYSTEM

Before we formulate a ration for dairy goats, it will be helpful to know something about the animal's digestive system. People familiar only with human diets and perhaps those of dogs and cats should especially examine the process of rumination, because goats are ruminants. Like cows and sheep, they have "four stomachs."

The process of rumination serves a very definite purpose, and has an important bearing on the dietary needs of the animal. Ruminants feed only on plant matter, which consists largely of cellulose and other carbohydrates and water, making adaptations in the structure and functioning of the stomach and intestines necessary. We commonly speak of "four stomachs," but in reality the large rumen (or paunch), the reticulum, and the omasum ("many plies") are all believed to be derived from the esophagus, while the fourth stomach, the abomasum or true stomach, corresponds to the single stomach of other mammals.

Vast numbers of protozoans and bacteria live in the rumen and reticulum. When food enters these "stomachs," the microbes begin to digest and ferment it, breaking down not only protein, starch, and fats, but cellulose as well. The larger, coarser material is periodically regurgitated as the cud, rechewed, and swallowed again. Eventually the products of the microbial action (and some of the microbes themselves) move into the "true" stomach where final digestion and absorption take place.

No mammal, including the goat, has cellulose-digesting enzymes of its own. Goats rely on the tiny animals in their digestive tracts to break down the cellulose in their herbiverous diet. You might say you're feeding the microbes and the microbes feed the goat, for without them the grass and hay would have no food value.

Let's back up a bit to take another look at these stomachs, for not only are they of obvious importance to the goat: the goat owner (or at least the kid raiser) has some control over their development.

Watch a newborn or very young kid sucking. She stretches her neck out to get her milk. Due to the stretching process the milk goes past a slit in the esophagus, bypassing the first two stomachs and ending up in the omasum. Here it is mixed with digesting fluids and is passed on to the fourth stomach, or abomasum.

Contrast this with a pan-fed kid, especially one fed only two or three times a day instead of four or five, and who is therefore more hungry and greedy. It must, first of all, bend down to drink rather than stretch upwards. Some of the milk slops through the slit in the food tube and falls into the first stomach, the rumen, where it doesn't belong. There is nothing else in this compartment, since milk is the only feed consumed. There is no bulk. Gas forms, and scours are likely to result.

ROUGHAGE AND THE RUMEN

The good goat raiser will strive to keep milk out of the rumen by proper feeding. Moreover, the breeder will work to develop the rumen and reticulum the way they should be developed by encouraging the kid to eat roughage at an early age. Here's why this is important.

A young milk-fed kid has about 30 percent of its stomach space occupied by the rumen and reticulum. At maturity, a well-developed doe has a rumen that occupies 80 percent of the stomach space and a reticulum that takes up 5 percent. (As an illustration of why feeding requirements of ruminants differ from single-stomach animals, note that a horse's stomach holds 12-19 quarts, while the four of a cow hold over 250 quarts!)

The rumen does not increase to this size without proper stretching or development. Early feeding of roughage is essential.

Now let's examine a mature doe. She takes little time to chew. Notice how she draws in her neck to swallow, allowing the food to slip through the slit in the esophagus to the rumen. A slight fermentation begins as the microbes go to work. When at leisure, the doe brings up some of this material by regurgitation and she "chews her cud." This time the mastication process is thorough. Now she extends her neck when she swallows, and the cud goes to the third stomach, or omasum.

Set a pail of water in front of the goat. Notice how she extends her neck to the far side to drink. This insures that the fluid goes to the omasum where it belongs, not to the rumen.

The goat must have a well-developed rumen to function properly, and a bulky diet to keep the rumen working properly. This means hay (or other roughage) forms the basis for the diet.

In his book *Goat Husbandry* (London: Faber and Faber, 1957), David Mackenzie maintains that bulk is necessary for good milk production. He points out that milk production in British goats dropped 20 percent in 15 years…and by 12 percent just in the four years following the "derationing" of animal feedstuffs in 1949.

The reason, he believes, is that when concentrates were rationed during the war years, the official concentrate ration for a milking goat was adequate if she had plenty of bulk food such as hay and roots. The allowance for kids and young stock was much more restrictive, and milk for kids very much so. His charts show a steady increase in milk production based on about 3,000 records from the British Goat Society, and a dramatic *decrease* after rationing was lifted. His conclusion was that excessive feeding of milk and concentrates to young goats prevents full development of the rumen.

The rumen, therefore, is of prime importance, along with the tiny animals that inhabit it. Next task: feeding those tiny animals.

They are conditioned to what they've been accustomed to eating. Change their diet and they can't cope. The result is a sick goat. Therefore, make any feed changes gradually. Many a goat raiser has fed a goat an armload of cornstalks salvaged from the garden after harvesting sweet corn, and when the goat gets sick or dies, the cornstalks get the blame. In reality the problem was one of overload. Feed such delicious things sparingly, along with the regular diet, and everybody—protozoans, bacteria, goat, and you—will be happier.

FEEDING FOR MILK PRODUCTION

In addition to the special needs of the goat relative to rumination, it's important to feed her as a *dairy* animal. Production of milk requires more protein than would be needed just for body maintenance, for example. So a milking doe is fed a ration of at least 16 percent protein, while a dry mature doe or buck will do well on 12 percent. Protein is expensive, and any excess is just wasted. You want to make sure the diet has enough, but not too much.

Dairy animals also have a greater need for calcium and certain trace minerals. (See page 63 for additional salt requirements for milking animals.)

NUTRITIONAL REQUIREMENTS

It would be very helpful to think in terms of minimum daily requirements for humans, which most of us are familiar with nowadays. Goats, too, have minimum daily requirements. Remember this and you'll be less tempted to stake the animal in a brush patch and assume she's "fed" just because she filled her belly. She has no more nourishment in that situation than you would if you lived on candy bars and soda pop.

Yet, there are people who treat goats like that…or toss them an armload of grass clipings or an ear of corn…or who mix up a concoction of corn and oats. The only difference is one of degree. (Incidentally, lawn clippings are rich in vitamin A. Some people cure them like hay, and goats love them. Of course, you wouldn't use clippings from any lawn that's treated with chemicals.)

Actually, there are no "minimum daily requirements" listed for goats. With several notable and recent exceptions, very little research has been done with goats, certainly in comparison with the more economically important livestock such as cattle and hogs. But we can look at nutrition in general and make some assumptions based on what is known about other ruminants.

WATER

All feeds contain water, organic matter, and mineral matter or ash.

Water is vital to life, of course, but it's also important in feed formulations because the quantity of water in various plants affects their place in the ration. Dry grain, for example, might contain 8 to 10 percent water. Green growing plants might contain 70 to 80 percent water. An animal fed the succulent plants ingests an enormous amount of water in order to get nutrients. Like any other livestock, however, goats should have free access to fresh water.

CARBOHYDRATES

Of the plants' dry matter, about 75 percent is carbohydrates, the chief source of heat and energy. These carbohydrates include sugars, starch, cellulose, and other compounds.

The sugars and starch are easily digested and have high feed value. Cellulose, lignin, and certain other carbohydrates are digested only with great difficulty and therefore it takes energy to digest them: their feed value is correspondingly lower. (This is one reason goat raisers prefer "fine-stemmed, leafy green hay." The fine stem means less lignin and hard-to-digest materials.)

If you buy feed, the feed tag on the sack will have the carbohydrates divided into two classes: crude fiber (or just plain fiber), and nitrogen-free extract. Nitrogen-free extract is the more soluble part of the carbohydrates, and includes starch, sugars, and the more soluble portions of the pentosans and other complex carbohydrates. It also includes lactic acid (found in milk) and acetic acid (in silage). Oddly enough, nitrogen-free extract also includes the lignin, which has a decidedly lower feeding value than cellulose.

FATS

Feed tags also list "fat," which actually includes fats and oils. They're the same except that fats are solid at ordinary temperatures while oils are liquid. In grains and seeds, fat is true fat. In hays and grasses, much fat consists of other substances. Many of these are vital for life, including cholesterol; ergosterol (which can form vitamin D); and carotene, which animals can convert into vitamin A.

PROTEINS

The proteins and other nitrogenous compounds are of outstanding importance in stock feeding: many discussions on goat feeding focus on protein to the exclusion of everything else.

Proteins are exceedingly complex, each molecule containing thousands of atoms. There are many kinds of proteins, some more valuable than others. (Livestock feeders speak of the "quality" of protein.) All are made up of amino acids, and protein must be broken down into amino acids before it can be absorbed and utilized by the body. There are at least twenty-four amino acids, but since they can combine like letters of the alphabet, there could be as many proteins as there are words in the dictionary.

The protein in plants is concentrated in rapidly growing parts (the leaves) and the reproductive parts (the fruits, or seeds). In animals, protein comprises most of the protoplasm in living cells and the cell walls, so it's important for muscles, internal organs, skin, wool or hair, and feathers or horns, and it's an important part of the skeleton.

Protein, or crude protein, includes all the nitrogenous compounds in feeds. It's of extreme importance to the animal caretaker: it's obviously essential for life, but needs vary among classes of animals. Protein requirements are higher for young and growing animals, reproduction, and lactation. And because protein is the most expensive portion of livestock feed, you won't want to offer more than necessary.

MINERALS

"Ash" indicates the mineral matter of the ingredients. Minerals in plants come from the soil, but the mineral content of animals is higher than that of plants. Calcium and phosphorus are particularly important since they are the chief minerals in bone and in the body. The body contains about twice as much calcium as phosphorus, and the proper balance is important.

Other minerals are needed only in trace amounts, but they are vital. Iodine, for example, prevents goiter; iron is important for hemoglobin, which carries oxygen to the blood; copper, which is a violent poison, is also a vital necessity in trace amounts, for a lack of iron, copper, or cobalt can result in nutritional anemia. Other trace minerals include potassium, magnesium, manganese, zinc, and sulfur.

TOTAL DIGESTIBLE NUTRIENTS

Net energy values of livestock feeds are expressed in *therms* instead of calories. Since a therm is the amount of heat required to raise the temperature of 1,000 kilograms of water one degree centigrade, one therm is equal to 1,000,000 calories.

Nutrients are constantly being oxidized in tissues to provide heat and energy. This oxidation maintains body heat and powers all muscular movements. Since the digestion of roughages requires more energy, it follows that one pound of Total Digestible Nutrients in roughages will be worth less than one pound of TDN in concentrates, which will not use up so much of its energy just being digested. TDN (Total Digestible Nutrients) refers to all of the digestible organic nutrients: protein, fiber, nitrogen-free extract, and fat. (Fat is multiplied by 2.25 because its energy value for animals is approximately 2.25 times that of protein or carbohydrates.)

"Digestible," of course, refers to nutrients that can be assimilated and used by the body. For this reason protein or crude protein is different from digestible protein. Digestible nutrients are determined in the laboratory by carefully measuring the amount of feed consumed and analyzing its content, and then analyzing the waste products. (Animal feces are largely undigested food, in contrast to human feces, which have a larger proportion of spent cells and other true "waste.")

VITAMINS

Another important consideration in feeding: vitamins. Vitamins were largely unknown before 1911, and there is still more to learn about them. But as of now, the only two of any consequence to goats are vitamins A and D.

Vitamin A is of prime importance to dairy goats because it's necessary for growth, reproduction, and milk. It is of less importance in maintenance rations. Vitamin A is synthesized by goats that receive carotene in their diets. The chief sources of carotene are yellow corn and leafy green hay. Common symptoms of vitamin A deficiency are poor growth, scours, head colds and nasal discharge, respiratory diseases including pneumonia, and blindness. A severe lack of vitamin A prevents reproduction or produces weak (or dead) young at birth.

The other important vitamin for goats is vitamin D. As with other animals, lack of this vitamin causes rickets, weak skeleton, impaired joints, and poor teeth. Vitamin D is necessary to enable the body to make proper use of calcium and phosphorus. The best and chief source is sunshine, but it is also available in sun-cured hay.

The B-complex vitamins are manufactured in the rumen; therefore, the feeder has no concern with them directly. Vitamin E seems to have no special application to goats. Vitamin C is synthesized. (Only humans, monkeys, and guinea pigs lack the ability to manufacture vitamin C.) Vitamin K is also synthesized.

With this very brief background we can begin to formulate a goat ration.

COMPOSITION OF GOAT FEEDS

The main tool we'll use will be a list of the protein content of the common goat feeds (see pages 60-61). The idea is to combine the various ingredients you have available in such a way that the combination will contain the desired amount of protein, or more accurately, digestible protein. But since protein is only one element of feed value, we must also keep in mind the minerals, vitamins, fiber, and palatability, as well as cost.

Because goats are ruminants, the main portion of their diet is roughage, so let's begin with that.

ROUGHAGE

Roughage can be green, growing plants, including grasses, clovers, and the trees and shrubs goats eat. It can also be plants in dried form, called hay. There are two types of hay: legume hay, made from alfalfa or clover; and carbonaceous hays made from timothy, brome, or other grasses. Corn stover (dry corn stalks), silage (fermented corn plants or hay plants), comfrey, sunflower and Jerusalem artichoke stems and leaves, and root crops such as mangel beets, Jerusalem artichokes, carrots, and turnips are also considered roughages.

Green forages are rich in most vitamins except D and B12. But if the animal is grazing it's getting sunshine and vitamin D, and ruminants can synthesize B12. Rapidly growing grass is also rich in protein.

However, because of the high water content of succulent green feed (and roots, too), these are low in minerals. The lack of minerals combined with the high water content (an animal could *drown* before it got enough nutrients from really lush grass) means that such forage does not constitute an adequate diet by itself. And lush forages can cause bloat.

Alfalfa or clover hay is considered the ideal for goats because of the high protein content and because these are rich in calcium, the most important mineral. Good alfalfa or clover hay is cut before full bloom, when its nutritive value is highest. It is sun-cured quickly. Rain or slow curing in damp weather leaches nutrients out of hay. Good hay is fine-stemmed, bright green, and leafy.

Most of the nutrition is in the leaves. Hay that is baled when it's too dry suffers much shattering and loss of leaves.

The importance of good hay can be illustrated by the fact that good alfalfa can have as much as 40 milligrams of carotene per pound, but alfalfa that is bleached and otherwise of poor quality can have as little as 4

AVERAGE COMPOSITION OF SELECTED GOAT FEEDS

	Crude Protein	Digestible Protein	Fat
Alfalfa hay	15.3%	10.9	1.9
Bermuda grass	7.1	3.6	1.8
Birdsfoot trefoil	14.2	9.8	2.1
Brome	10.4	5.3	2.1
Red clover	12.0	7.2	2.5
Mixed grass	7.0	3.5	2.5
Johnson grass	6.5	2.9	2.0
Soybean, early bloom	16.7	12.0	3.3
Timothy, early bloom	7.6	4.2	2.3
Succulents			
Green alfalfa, early bloom	4.6	3.6	0.7
Bermuda grass pasture	2.8	2.0	0.5
Cabbage	1.4	1.1	0.2
Carrot roots	1.2	0.9	0.2
Kale	2.4	1.9	0.5
Kohlrabi	2.0	1.5	0.1
Mangel beets	1.3	0.9	0.1
Parsnips	1.7	1.2	0.4
Potatoes	2.2	1.3	0.1
Pumpkins (with seeds)	1.0	1.3	1.0
Rutabagas	1.3	1.0	0.2
Sunflowers (entire plant)	1.4	0.8	0.7
Tomatoes (fruit)	0.9	0.6	0.4
Turnips	1.3	0.9	0.2
Barley	12.7	10.0	1.9
Steamed bone meal	7.5	—	1.2
Buckwheat	10.3	7.4	2.3
#2 dent corn	8.7	6.7	3.9
Linseed meal	35.1	30.5	4.5
Cane molasses	3.0	—	—
Oats	12.0	9.4	4.6
Field peas	23.4	20.1	1.2
Pumpkin seed	17.6	14.8	20.6
Rye	12.6	10.0	1.7
Soybeans	37.9	33.7	18.0
Sunflower seed, w. hulls	16.8	13.9	25.9
Wheat (average)	13.2	11.1	1.9
Wheat bran	16.4	13.3	4.5

Fiber	Nitrogen-Free Extract	Mineral Matter	Calcium	Phosphorus
28.6	36.7	8.0	1.47	0.24
25.9	48.7	7.0	0.37	0.19
27.0	41.9	6.0	1.60	0.20
28.2	39.9	8.2	0.42	0.19
27.1	40.3	6.4	1.28	0.20
30.9	43.1	6.5	0.48	0.21
30.5	43.7	7.5	0.87	0.26
20.6	37.8	9.6	1.29	0.34
30.1	44.3	4.7	0.41	0.21
5.8	9.3	2.1	0.53	0.07
6.4	12.2	3.1	0.14	0.05
0.9	4.4	0.7	0.05	0.03
1.1	8.2	1.2	0.05	0.04
1.6	5.5	1.8	0.19	0.06
1.3	4.3	1.3	0.08	0.07
0.8	6.0	1.0	0.02	0.02
1.3	11.9	1.3	0.06	0.08
0.4	17.4	1.1	0.01	0.05
1.6	5.2	0.9	—	0.04
1.4	7.2	1.0	0.05	0.03
5.2	7.9	1.7	0.29	0.04
0.6	3.3	0.5	0.01	0.03
1.1	5.8	0.9	0.06	0.02
5.4	66.6	2.8	0.06	0.40
1.5	3.2	82.1	30.14	14.53
10.7	62.8	1.9	0.09	0.31
2.0	69.2	1.2	0.02	0.27
9.0	36.7	5.7	0.41	0.85
—	61.7	8.6	0.66	0.08
11.0	58.6	4.0	0.09	0.33
6.1	57.0	3.0	0.17	0.50
10.8	4.1	1.9	—	—
2.4	70.9	1.9	0.10	0.33
5.0	24.5	4.6	0.25	0.59
29.0	18.8	3.1	0.17	0.52
2.6	69.9	1.9	0.04	0.39
10.0	53.1	6.1	0.13	1.29

milligrams per pound. Poor hay may be difficult to distinguish from some straw, which is the plant residue left after the grain—commonly oats, wheat, or barley—has matured and has been harvested. Straw contains much fiber, especially lignin, and is used as stock feed only in dire emergencies. It has no place in a dairy goat feeding program. Straw is bedding.

The carbonaceous hays have less protein and less calcium than the legumes, and these deficiencies must be made up in the concentrate or grain ration.

Other hay plants include barley (cut when the seed heads are in the immature "boot" stage), birdsfoot trefoil, Bermuda grass, lespedeza, marsh or prairie grasses, oat or wheat grasses, soybeans, or combinations of these. For the benefit of inexperienced farmers, it will be well to point out again that hay is made by cutting green growing plants and drying or "curing" them in the sun. Wheat, barley, oats, and soybeans, for example, can be cut when young for hay. If allowed to mature, the nutriment goes into the grain and the stems and leaves become yellow and have little food value: the plant that could have been hay becomes straw.

Good alfalfa has about 13 percent protein; timothy and brome are usually closer to 5 percent.

While roughages are the most important part of the diet of a ruminant, they alone don't provide all of the needed vitamins and minerals, nor do they provide sufficient energy. Alfalfa hay has about 40 therms (energy) per 100 pounds; corn and barley have twice that. Especially if carbonaceous hays are fed (5 percent protein), additional protein and calcium are required. Hays do not provide sufficient phosphorus. These missing elements are provided in the concentrate ration.

A mature goat will require anywhere from three to ten pounds of hay per day depending on the type, quality, waste, and other factors.

The concentrate ration is often called the grain ration, but this can be misleading. Here's why.

For lactating animals, the protein content of the concentrate ration should be about 16 percent if the roughage is a good legume. With less protein in hay, more must be added to the concentrate. For dry does, 12 percent protein is sufficient.

Corn has about 9 percent protein and only 6.7 percent digestible protein. Oats has about 13 percent protein and 9.4 percent digestible protein. Therefore, a mixture of equal parts of corn and oats would contain 11 percent protein or about 8 percent digestible protein. Clearly, these grains alone will not meet the demands of the growing or milking animal. Protein supplements in the form of soybean oil meal (sometimes listed on feed tags as SOM), or linseed or cottonseed oil meal, must be added.

Milking animals also require more salt than is needed for animals on maintenance rations. It is usually added at the rate of 1 pound per 100 pounds of feed.

Because of the need for bulk in the diet of a ruminant, a concentrate ration should not weigh more than 1 pound per quart. Bran is most commonly used for bulk. (Beet pulp is sometimes used for does, but extended feeding of beet pulp to bucks can cause urinary calculi.) The weight of grain varies with quality, but this chart shows some of the averages you can expect to find.

WEIGHTS OF SOME COMMON GOAT FEEDS	
Barley, whole	1.5 pounds per quart
Buckwheat, whole	1.4
Corn, whole	1.7
Linseed meal	0.9
Molasses	3.0
Oats	1.0
Soybeans	1.8
Sunflowers seeds	1.5
Wheat, whole	1.9
Wheat bran	0.5

And finally, since goats generally shun dusty ground feed such as that normally fed to cows, the grains should be crimped or cracked or even whole, rather than ground to flour. Cows do not digest whole grains well. Whole corn goes in one end and out the other. Goats seem to have better powers of digestion.

This is fine for the goat owner who wants to mix a ration rather than buy bagged feed because it eliminates the bother and expense of grinding. But it also means the fine ingredients—salt, bran, oil meal, and minerals—can't be mixed into the grain. They sift to the bottom and the goats won't consume them. To overcome this, most goat feeds contain cane molasses. In addition to binding the ingredients, molasses makes the feed less dusty, is an important source of iron and other important minerals, it increases the palatibility of the feed, and does fed ample molasses during gestation are less likely to encounter ketosis. Molasses contains about 3 percent protein, but none of it is digestible.

There is some evidence, at least in dairy cows, that excess molasses interferes with the digestibility of other feeds. The digestive processes attack the more easily assimilated sugars in molasses to the detriment of other feedstuffs. Even so, molasses is an important feed for goats.

FORMULATING A RATION

At last we're ready to formulate a ration. Here are several base formulas to work from.

1: Ration for a milking doe fed good alfalfa hay.
12.6% digestible protein.

Corn	31 pounds
Oats	25
Wheat bran	11
Linseed oil meal	22
Cane molasses	10
Salt	1

2: Ration for a milking doe fed good alfalfa hay.
12.6% digestible protein.

Barley	40 pounds
Oats	28
Wheat bran	10
Soybean oil meal	11
Cane molasses	10
Salt	1

3: Rations for a milking doe fed non-legume hay.
21.2% digestible protein.

Corn	11 pounds
Oats	10
Wheat bran	10
Corn gluten feed	30
Soybean oil meal	24
Cane molasses	10
Salt	1

4: Ration for a milking doe fed non-legume hay.
   21.2% digestible protein.
Barley	25 pounds
Oats	20
Wheat bran	10
Soybean oil meal	25
Linseed oil meal	15
Salt	1

5: Ration for dry does and bucks.
   9.8% digestible protein.
Corn	58 pounds
Oats	25
Wheat bran	11
Soybean oil meal	5
Salt	1

6: Ration for dry does and bucks.
   10.1% digestible protein.
Barley or wheat	51.5 pounds
Oats	35
Wheat bran	12.5
Salt	1

These rations, followed more or less faithfully, could be expected to produce good results. There will be minor variations because the feed value of grains depends in part on variety, weather, and the fertility of the soil that produced them. Most grains grown in the Pacific Northwest are lower in protein than the same grains grown elsewhere; old-fashioned, open-pollinated corn has more protein than the hybrids in common use today, and so on.

There are more serious considerations than these, however. One is that certain ingredients might not be available in your locale, or others may be more common and therefore less expensive than those listed. Grains can be substituted for one another by using the chart showing protein contents.

You can determine the weight of protein in a given feed ingredient, and by working with batches of 100 pounds, merely move the decimal point two places to the left to get the percentage of protein in a ration.

As an example, let's look at a small homestead farm that produces its own grain. The previous year's corn crop was almost a total failure due to a wet spring, summer drought, and early frost. But other grains were available. Here's what the milking does were fed:

Feed	Weight	% Crude Protein	% Digestible Protein	Pounds of Protein
Soybeans	20 pounds	37.9	33.7	6.74
Barley	19	12.7	10.0	1.9
Oats	20	13	9.4	1.88
Buckwheat	5	10	7.4	.37
Wheat bran	5	16.4	13.3	.66
Corn	10	9	6.7	.67
Linseed meal	10	34	30.6	3.06
Molasses	10	3	0	0
Salt	1	0	0	0
Total	100			15.28

Divide the pounds of protein (15.28) into the total weight of the ration (100 pounds). This feed has 15.28 percent protein.

It should be noted that some rations you will find elsewhere work with crude protein rather than digestible protein. Since there are no digestibility trials on goats, both methods have flaws. It may be easier to obtain figures on crude protein for locally grown feeds from your extension office, in which case all the ingredients should be calculated on the basis of crude protein.

This leads us to another—perhaps the most important—reason why every goat owner should have at least a basic knowledge of feed formulations. Goatkeepers are notorious for dishing out treats or making use of "waste." These are both admirable traits, but look what happens.

Assume a goat is receiving one pound of a commercial 16 percent (crude) mixture. Maybe it costs the owner $16 per hundredweight, and he can get corn for half that, or he grew a little corn for the chickens and has some extra. Or the goat just seems to "like" corn! So he decides to give the goat one-half pound of the regular ration and one-half pound corn.

Here's what happens:

Goat feed	50 pounds	8.0 pounds protein
Corn	50 pounds	4.5 pounds protein
	Total	12.5 pounds protein

That 16 percent mixture drops to 12.5 percent protein. That might be enough for the goat to maintain her own body, but not to produce kids and milk.

The same thing happens when the animal is given garden "waste" or trimmings. Such fodder replaces roughage, not grain, but even then it can cause unbalancing of the diet because elements of hay, for instance, will be missing from most of the garden produce.

This is not to say that rations can't be manipulated or that the goat breeder shouldn't make use of what's available or cheap. It must be done with a certain amount of knowledge and discretion, however.

With this principle firmly in mind, let's examine some of the common feeds small farmers have available and show an interest in.

Soybeans deserve special mention because many people look at the price of the oil meals and wonder why the beans can't be fed whole. They can, with certain restrictions.

SOYBEANS

Soybeans contain what is called an "antitrypsin factor." Trypsin is an enzyme in the pancreatic juice that helps produce more thorough decomposition of protein substances. The antitrypsin factor doesn't let the trypsin do its job, which means the extra protein in the soybeans is lost, not digested. The antitrypsin factor can be destroyed by cooking and it isn't present in soybean oil meal.

However, rumen organisms apparently inactivate the antitrypsin factor when raw soybeans are fed in small amounts. Current recommendations for dairy cattle are that the ration not contain more than 20 percent raw soybeans.

DO NOT FEED RAW SOYBEANS IF YOUR FEED CONTAINS UREA! The result will probably be a dead goat. Urea is not recommended for goats in any case, but many dairy feeds for cows contain it. So does LPS (Liquid Protein Supplement), which some feed dealers will try to sell you when you ask for molasses. Urea is a non-protein substance that can be converted to protein by ruminants, and some people do feed it to goats because it's less expensive than the oil meal protein supplements, but other goatkeepers have reported breeding problems with animals fed urea.

GARDEN PRODUCE

Most goats are raised on small farms or homesteads where grain and hay are not produced. Such places can still grow a great deal of goat feed if the basic principles of feeding are followed. You can "grow milk in your garden" by planting sunflowers (the seeds are high in protein and the goats will eat the entire plants), mangel beets, Jerusalem artichokes, pumpkins, comfrey, carrots, kale, turnips, and others. In addition, such "waste" as cull carrots and apples and sweet corn husks and stalks can be utilized in the goat yard. These are treated like pasture or silage: they replace part of the grain ration, but not all of it. Feed at least 1 pound of concentrates per head per day to milking animals.

WEEDS

Many people with more time than money and a keen interest in nutrition are avid collectors of weeds for their animals. For example, dandelion greens are extremely rich in vitamin A, and nettles are high in vitamins A and C. Goats relish these and other common weeds. It's just about impossible to imagine a real farmer on his knees gathering dandelion greens for his livestock...but many a goat farmer can, and reaps healthier animals and lower feed bills.

This brings up a point of particular interest to those who want to mix their own feeds because they aren't satisfied with commercially prepared rations. No one plant has everything any animal needs for nutrition. Goats seem to enjoy variety more than most domestic animals. Many goatkeepers prefer to provide food from as many different plant sources as possible to enhance the possibility that their animals are getting the nutrition they need naturally, without synthetic additives. They like grain mixtures of at least five or six different ingredients.

This isn't as "efficient" as modern agricultural methods. Farmers know that alfalfa is rich in protein and calcium, both important to dairy animals. A great deal of feed can be harvested from one acre of alfalfa, and alfalfa hay has become the norm. There are even herbicides that kill weeds in alfalfa to keep stands pure.

But almost any weed in your garden has more cobalt than alfalfa, and cobalt is required by ruminants to provide the bacteria in the digestive tract with the raw material from which to synthesize vitamin B12. Some, if not all, internal parasites rob their hosts of this vitamin.

Alfalfa (and clover) has little cobalt because lime in the soil depresses the uptake of this mineral, and lime is necessary for the growth (and the calcium content) of alfalfa. Agribusiness has found it more efficient to strive for high yields of alfalfa and then add the trace minerals to the concentrate ration. Homesteaders who don't mind gathering "weeds" can meet their animals' nutritional needs naturally...and without the cash outlay required for commercial additives.

Organic farmers have known this for years, of course, but when their beliefs were confirmed by scientists in 1974, the idea was hailed as "revolutionary." Researchers at the University of Minnesota compared nutritive value and palatability of four grassy weeds and eight broadleaf weeds with alfalfa and oats as a feed for sheep, which have roughly the same requirements as goats.

Lamb's-quarters, ragweed, redroot pigweed, velvetleaf, and barnyard grass all were found to be as digestible as alfalfa and more so than oat forage. All five weeds had more crude protein than oats and four had as much as alfalfa. Eight were as palatable as oat forage.

One caution about weeds, though: Don't gather them from along roadsides where the lead content may be high due to auto exhausts, and certainly not where spraying is done.

Also, under certain conditions, some weeds such as lamb's-quarters and pigweed (and even "normal" crops such as oat and wheat grass and sudan) can be toxic. When they are very young or when they grow rapidly after a setback such as drought, they can be dangerous. Some plants are also hazardous after being killed by frost. Your county agent can give you more specific information on these plants and conditions prevalent in your area.

Tree trimmings fall into the category of weeds. Tree leaves and bark are rich sources of minerals brought from deep within the earth by tree roots. Although goats love pine boughs, there have been reports that pine needles have caused abortions, so caution is advised. They are rich in vitamin C, although goats have no particular need for the vitamin, being able to manufacture it themselves. Also avoid wild cherry, since when the leaves are wilted they are poisonous. Some weeds are poisonous too, of course, including milkweed, locoweed, and bracken. Since these vary so widely in distribution, consult your county agent to see what's considered dangerous for cows in your area.

COMFREY

One more plant deserves special attention, because so many people are interested in it and because it's controversial. That's comfrey, or boneset.

There was a rash of statements from county agents and state departments of agriculture a few years ago knocking comfrey. Some of their reasons for not growing it are practical...for large farmers, not homesteaders. And some of their information is just plain wrong. There are many goat and rabbit raisers who swear by comfrey as a feed, a tonic, and as medication for certain conditions such as scours. They can't all be wrong.

Even aside from that, comfrey should be in every goat owner's garden. It is high in protein, ranking with alfalfa, although there is some question about the digestibility of the protein. But it is easier to grow and harvest than alfalfa, using hand methods. It is an attractive plant that even can be used for borders or other decorative applications: grow goat food in your front yard or flower bed! It has tremendous yields since it begins growing early in spring and grows back quickly after cutting. And it is a perennial. It can be dried for hay, although that entails a lot of work because of the thick stems. It must be cured in small amounts on racks rather than left lying on the ground.

MINERALS

No discussion of goat feeds would be complete without mentioning minerals. Most goat raisers supply mineralized salt blocks free choice, and also add dairy minerals to the feed.

While there is no sound research into the matter, there is some indication that this is unnecessary, expensive, and perhaps even dangerous. Too much of a good thing can be as bad as too little. Goats that are well fed on plants and plant products from a variety of sources, grown on organically fertile soil, probably have little or no need for additional minerals.

There are certain exceptions. Plants grown in the goiter belt (from the Great Lakes westward) are low in iodine, so iodized salt will be good insurance. Certain areas of Florida, Maine, New Hampshire, Michigan, New York, and Wisconsin and western Canada have soils deficient in cobalt. Parts of Florida are deficient in calcium.

Selenium deficiency can cause white muscle disease (see Chapter 8). Most soils in the central and eastern U.S. (and some in other areas) are deficient in selenium. Injecting does with selenium 15 to 30 days before kidding and kids at three to four weeks of age will prevent this serious disease. There's a relationship between selenium and vitamin E: both are usually administered at the same time.

On the other hand, some soils, and the feeds grown on them, are high enough in selenium to cause poisoning.

Phosphorus is a vital ingredient of the chief protein in the nuclei of all body cells. It is also part of other proteins, such as the casein of milk. Therefore, it is of extreme importance to growing animals producing bone and muscle; pregnant animals, which must digest the nutrient needs for the growing fetus; and for lactating animals, which excrete great quantities of these minerals in their milk. Vitamin D is required to assimilate calcium and phosphorus. Also, the ration of calcium to phosphorus is critical: it should be 1.5:1. When legume forages are fed, the goat might need more phosphorus in the concentrate ration to maintain the 1.5:1 balance; when carbonaceous forages are used, supplemental calcium might be required. Roughages, especially legumes, are high in calcium, and grains are high in phosphorus. If these crops are grown on soils rich in these minerals, the well-fed goat is likely to get enough of them.

A goat lacking phosphorus will show a lack of appetite, it will fail to grow or will drop in milk production if in lactation, and it may acquire a depraved appetite such as eating dirt or gnawing on bones or wood. (Many goats like to chew on wood even without a phosphorus deficiency.) In extreme cases stiffness of joints and fragile bones may result.

However, overfeeding calcium can be dangerous, too, especially for young animals. Lameness and bone problems can result later from excess calcium.

Iron is 0.01 to 0.03 percent of the body, and is vital for the role it plays in hemoglobin, which carries oxygen in the blood.

Copper requirements are about one-tenth those of iron, and in greater amounts copper is a deadly poison.

Nutritional anemia can result from lack of iron, copper, or cobalt. (This is different from pernicious anemia in humans.) But it's very rare.

Other trace minerals are potassium, magnesium, zinc, and sulfur. If you feed your goats well, a trace mineral salt block will last a long time. In that case there is no need for adding minerals to the feed.

It can be seen that feeding is a science...and an art. Goats are not "hayburners" or mere machines to be fueled haphazardly. You wouldn't burn kerosene in a high-powered sports car, and you can't get the full potential from a goat fed improperly.

Summing up: Feed your goats 1 pound of concentrate for maintenance and 1 pound extra for each 2 pounds of milk produced, along with all the hay they will eat. Some will do better on less, others will want more: that's the art, or part of it. "The eye of the master fattens the livestock."

Remember these basic concepts:
- The ration should come from as many different sources as possible, and fertile soil.
- Avoid sudden changes in feed, which may result in overloading the rumen bacteria and microbes.
- Pay attention to protein levels as well as vitamin and mineral content of the plants and grains you feed.
- Treat each animal as an individual, for they have different needs according to age, condition, production, and personal quirks.

CHAPTER 7

Grooming

Goats require a minimum of care...but that doesn't mean they require *no* care. The goat owner will quickly learn how to disbud, tattoo, clip, and trim hooves.

Let's begin with the most important: hoof trimming.

HOOF CARE

The horny outside layer of the goat's hoof grows much like your finger-nails and must be cut off periodically. Gross neglect of this duty can cripple the goat. But it's a simple job, and for the small herd it won't take more than a few minutes a month.

How often you trim hooves depends on several factors. At some times hooves grow faster than at other times, or there are differences among animals. Goats living on soft spongy bedding will need more hoof attention than goats that clamber on rocks. (One goat book claims that if a good-sized rock is placed in the goat pen, the animals will stand on it and keep their hooves worn down. It sounds logical, but everyone I know who has tried it says it doesn't work.)

TOOLS

There are several methods of trimming, requiring different tools. The simplest trimming tool is a good sharp jackknife. Some people prefer a linoleum or roofing knife, and others swear by the pruning shears (the kind you use for your roses). However, I don't think anyone who has used an honest-to-goodness goat hoof trimmer would ever want to go back to the more primitive tools. (Some catalogs selling sheep equipment call them "foot rot shears," a rather ugly and misleading name for the same tool.)

For light trimming and finishing, many people like a Surform (a small woodworker's plane with blades much like a vegetable grater).

METHODS

Hoof trimming is easier if you have a helper, or if your milking stand is of a type that allows you to lock the goat in and still have room to get at all four feet.

With the goat secured, stand against her rear (tail-to-tail, as it were), grasp one hind leg and lift it up between your legs. Some goats don't seem to mind such acrobatics; others will protest rather violently. Keep a firm grip on the ankle and be exceedingly careful with the knife or other tool so that in case she does kick, you won't be injured. This is one reason many people prefer the shears, which is much safer than a knife. In either case, it's a good idea to wear heavy gloves.

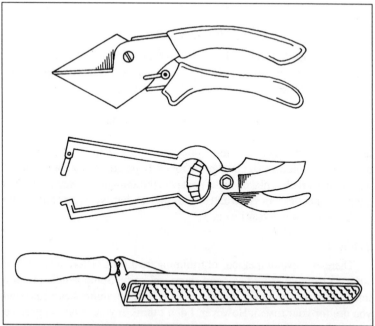

Three tools for trimming: pruning shears (top), goat hoof trimmers (middle), and Surform (below).

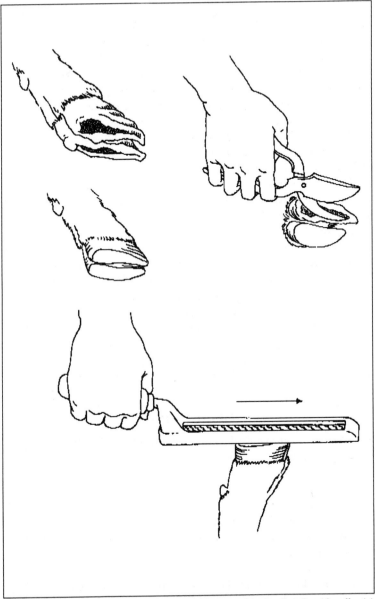

Trimming the hooves: (a) the overgrown goat hooves; (b) trimming the sidewalls; (c) properly trimmed hooves; (d) using the Surform for finishing.

With the point of the tool, clean out all the manure and dirt imbedded in the hoof. If the hoof has not been trimmed for some time it will have grown underneath the foot and can contain quite a lot of crud.

Next trim off those folded-over edges and obviously excess sidewalls. (At this point you might notice that dry hooves can be very hard. They're easier to trim after the goat has been walking in wet grass, and the shears, which offers leverage, cuts hard hooves more easily than a knife.)

Excess heel and toes should be trimmed off.

Then—here's where the Surform comes in handy—carefully take thin slices from the entire bottom of the foot. The portion near the toe invariably needs more of this finishing work than the heel. You can cut quite safely until the white portions within the hoof walls look pinkish.

Let the goat stand on the hoof and see how it looks. A goat with good hooves stands squarely. A kid a few weeks old has the ideal hoof you're aiming for.

Do the other hind hoof the same way. Then squat down beside a front leg, bring the foot up over your knee, and repeat the process.

In extremely bad cases where the goat looks like it's wearing pointed elf's shoes, it may take several trimmings to get them back in shape. In such cases it's better not to cut too much at once...and if they're really bad, the goat isn't likely to stand around patiently while you finish anyway. Take off as much as you can and come back a day or two later.

Bucks have hooves, too! These poor fellows are more likely to be neglected than the girls are, but of course they shouldn't be.

DISBUDDING

The other major grooming duty for goat owners is disbudding. Much has been written about the advantages and disadvantages of horns, and even more has been said around goat barns. Here's a summary of the arguments on both sides.

For horns:

Horns are protection against dogs and other predators; horns are beautiful and natural and a goat doesn't look like a goat without them; disbudding is ghastly.

Against horns:

They are dangerous to other goats and to people, especially children; it's impossible to build a decent manger that will accomodate a nice set of horns; horns are a disqualification on show animals (except for Pygmies);

horns aren't really much protection against dogs—look to better facilities instead.

One thing is certain: disbudding is *much* better, and easier, than dehorning.

Disbudding involves destroying the horn bud on a very young animal, before the horns really start to grow. Dehorning, on the other hand, is the surgical removal of grown or growing horns. Dehorning can be quite painful and even dangerous to the goat, and so upsetting to the surgeon that even many trained vets won't do it, or at least not more than once. They certainly don't solicit the business.

Disbudding is relatively quick, easy, and painless, although it might not appear so to the neophyte.

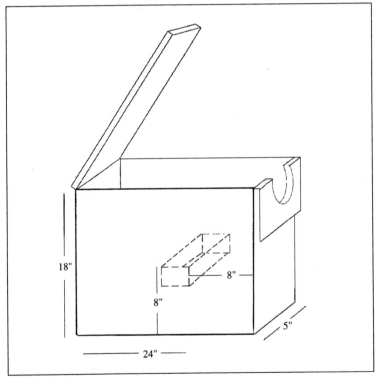

This kid holding box, made of ¾" plywood, keeps the kid still for finishing.

THE DISBUDDING IRON

The recommended tool is the electric disbudding iron. Kid-size disbudding irons are available, or you can make one from a large soldering iron (with a point about the size of a nickel). The point must be ground flat. Get the iron hot enough to "brand" a piece of wood with little pressure. Hold the kid on your lap. (If you'll be doing a lot of disbudding, you'll eventually want to construct a kid holding box, which is also handy for tattoing and other tasks.) If the horn has not yet erupted or you're not too sure of yourself, trim the hair around the horn button with a small scissors.

Then, holding the kid firmly by the muzzle, press the hot iron into the button and hold it there to a count of fifteen.

There will be acrid smoke from burning hair, violent struggling (which isn't too violent with a kid weighing 8-12 pounds), and maybe some screaming. But when the fifteen seconds are up, everything will be back to normal...except maybe your heartbeat.

Console the kid and compose yourself while the iron heats up again, then do the other horn button.

After it's all over, offer the kid a bottle of warm milk and she'll forget all about it. And remind yourself that the next one will be easier.

I knew one hardy, homesteader-type lady who didn't have electricity and who heated a metal rod in her woodburning stove for disbudding.

CAUSTIC

Another method, less hair-raising but also less successful and potentially more dangerous, is to burn the horn buds with dehorning paste, which is a caustic. Several types and brands are available from farm supply stores and mail order houses.

With this method, the hair is clipped around the button and Vaseline is applied *around,* not over, the area. Then the caustic is applied. Kids treated in this manner should be isolated for half an hour (one lady holds the kid on her lap while she watches television) so no other kids will lick at it and so it doesn't rub the caustic and get it on other kids or other parts of its body. The stuff can cause blindness if it gets in the eyes, and it will be quite painful on other parts of the body.

Kids' horns grow much sooner than calves': the directions on caustic were written for calves. Ignore them, as regards the time to do the job. The best time to disbud a goat is when it's a few days old.

Caustic might seem easier or less traumatic than the hot iron for the person performing the operation, and the hot iron might *seem* to be cruel

Disbudding horns by using caustic potash. From upper left: Clip hair around horn button. Then cut two peices of adhesive tape large enough to cover the horn button. Apply Vaseline around the adhesive. Finally, remove adhesive and apply caustic.

and unusual punishment for the kid. In reality, the iron is more humane. Allowing the horns to grow might well be the cruelest alternative, if those horns some day tear a gash in another animal or put out an eye.

Some people observe that not all goats have horns: some are naturally hornless, or polled. They ask, why not breed for hornlessness?

One of the main reasons has been the genetic link between hornlessness and hermaphroditism: many goats born of hornless-to-hornless matings are hermaphrodites, or of both sexes, which from a practical standpoint means they are sexless. Horned (or disbudded) goats can produce polled offspring, but because disbudding generally takes place before the horns erupt, these naturally hornless kids are usually disbudded.

Bucks have more stubborn horn buds than do does, and there is also a difference among breeds: some people say Nubians have tougher horn buds. If scurs start to develop, merely heat up the iron and do the job over again. In some ways, scurs (thin, misshapen horns) are more dangerous and troublesome than horns. They can curve around grotesquely and grow into an animal's head or eye, and thin ones will be broken off repeatedly, resulting in pain and loss of blood.

DEHORNING

In some cases it might be necessary to dehorn a goat. For example, if you have a herd of hornless goats and bring in a new animal with horns, she's sure to cause problems.

Grown horns can be sawed off, usually with a special wire blade. They must be removed close to the skull, actually taking a thin slice of the skull with it, or the horns will grow back. There will be a great deal of blood, and obviously a mature animal is more difficult to control. An anesthetic will be required. This is a job for a vet, and as mentioned, most of them don't want to tackle it.

Again there is an alternative that, while it might sound easier and more humane, actually isn't.

If very strong rubber bands are placed tightly around the base of the horn, the horn will atrophy and fall off. Some people file a notch in the horn very close to the skull to keep the band down where it belongs, and others claim putting tape over the band holds it on. In any case, check the rubber bands regularly to make sure they haven't broken or moved. Some recommend scrubbing the horn area and the bands with iodine first to prevent tetanus.

The problems arise when the horn structure begins to weaken. A goat may butt another, or merely get the horn caught in a manger or other obstacle, and break it off. If it's not really ready to fall off there will be considerable pain and a great deal of bleeding. (Blood-stopping powder is a good thing to have in your barn medicine cabinet in any event. If you don't have any, a handful of cobwebs will do the job in an emergency.)

With proper management — perhaps isolation of the treated animal, the removal of all obstructions, and very frequent inspection—the rubber band method seems preferable over sawing, in most cases. It may take a month or so for the horns to fall off, depending on the strength of the bands and the horns. But disbudding the young kid is far better, for all concerned.

TATTOOING

If you raise registered animals, you'll have to tattoo them. If your goats aren't registered, tattooing is still a good idea. Tattoos are permanent identification numbers that can help your recordkeeping, they can identify a goat long after you sell it, and in some cases they have helped retrieve lost or stolen animals.

You'll need a tattoo set, available from farm supply stores, mail order houses, and small- animal equipment dealers. Get a 1/4- or 5/16-inch die. Use green ink: it shows up even on dark-colored animals.

You'll need a helper or a kid holding box. Older animals can be fastened in a stanchion or milking stand, but goats should be tattooed soon after birth.

First, clean the area to be tattooed with a piece of cotton dipped in alcohol or carbon tetrachloride. La Manchas are tattooed in the tail web: other goats are tattooed in the ears. Stay away from warts, freckles, and veins. (I like this story: "Tattoo in the ears!" people asked Judy Kapture, *Countryside* magazine's goat editor. "Doesn't that hurt?" She just waggled her dangling, pierced earrings at them and smiled!)

Next, smear a generous quantity of tattoo ink over the area. Paste ink can be applied from the tube; liquid ink requires a small brush such as a toothbrush.

Then place the tattoo tongs in position and puncture the skin with a firm, quick squeeze. (Be sure to test the numbers on a piece of paper first: they're backwards, like printer's type.) On very thin-eared kids the needles might go all the way through the ear. Gently pull the skin free. Some tattoo outfits have an ear release feature that eliminates this problem.

Now put some more ink on the tattoo and rub it in thoroughly with the small brush.

It will take about a month for the tattoo to heal thoroughly.

Tattoos can record a great deal of information, including the animal's age, which is indicated by a letter of the alphabet signifying the year of birth, as designated by the breed associations. (The letter for 1988 was Z; 1989 started over again with A. G, I, O, Q, and U aren't used.)

Registered herds have herd tattoo IDs assigned by the registries, and many breeders add another digit to number each goat in the order in which they arrived in a given year. Some people develop their own codes so they can tell at a glance, even years later, which family an individual belongs to, and even who the dam and sire were.

If you have trouble reading a tattoo, try holding a flashlight behind the ear in a darkened building.

TRIMMING

Very shaggy goats should have their hair clipped, especially around the udder. It's a good idea to confine trimming to the udder region in the winter to keep your goat warmer and your milk cleaner. The entire animal can be clipped in the spring to keep it cleaner and cooler and to discourage parasites. Just like a haircut, a nice clipping will greatly enhance the appearance of a show goat.

Electric clippers are nice, but hand clippers (made for dogs) work well. If you only have one or two goats and don't want to invest in clippers and can't borrow one, scissors will do.

CASTRATING

Maybe it isn't "grooming," but another essential task is castration. Buck kids saved for meat should be castrated by the time they're a few weeks old, and no buck kept or sold as a pet should be left unaltered. A three-month-old buckling is capable of breeding his sister, and a grown, unaltered buck is definitely not the ideal pet.

BURDIZZO

The most highly recommended method of making a buck a wether is the Burdizzo, a tool that crushes the cords to the testicles. It's simple, quick, bloodless, and sure.

KNIFE

For castrating with a knife, have a helper hold the kid by the hind legs, his back to the helper's chest. Make a quick, clean incision with a sharp knife, grasp the testicle, and pull it out. Remove the other one, and spray the wounds with antiseptic.

ELASTRATOR

It's also possible to castrate with rubber bands, although some people consider this method cruel. A special tool called an elastrator is available from farm supply houses. It looks like a pliers, but when you squeeze the handles the jaws open instead of close. The jaws have four prongs which hold and stretch very small but strong rubber bands. Place a band on the prongs; squeeze the handles to enlarge the band; place it into position above the scrotum (making certain the testes are in the sac and not *above* the band); and remove it. Obviously this causes the buckling some discomfort.

The testicle will atrophy and the band will fall off in several weeks.

—————————————— CHAPTER 8 ——————————————

Health Care

Goats are among the healthiest and hardiest of domestic animals. Most people who pay attention to proper feeding and other management details have very few health problems with their goats. In spite of this, some people seem to be overly concerned by disease and sickness, in my view.

My own views are different, and I'll explain my attitude so you'll be aware of my bias. Then, if you don't agree, perhaps you can find a more technical manual, written by a veterinarian or someone who shares your interest in sickness (the *Merck Veterinary Manual* is a standard reference in this field).

Sickness is only an absence of health; health is the natural state. If your animals get sick, it's because of wrong conditions of feed, environment, or in some cases breeding. Treating the symptoms will help in the short run, sometimes, but unless the underlying causes are corrected, time and money spent on medication is wasted.

What's worse, many "illnesses" have purposes, and by "curing" them we're compounding the problem. Scours, or diarrhea, is one example. It's fairly common in kids, and can result from feeding too much milk, feeding cold milk (when the kid isn't used to it), using dirty utensils, etc. You don't want to stop the diarrhea cold because that's nature's way of getting rid of the toxins. So you let it take its course while removing the *cause:* the excess milk, the cold milk, or the unclean utensils. Similarly, a completely worm-free goat is a near-impossibility, and not a desirable goal under any circumstances: the use of so many different vermifuges would do more harm than good.

We have been led to believe that germs are bad per se. Nothing could be further from the truth. Even most pathogenic organisms will have little or no effect on a healthy body; only when the host is weakened because of some other factor—such as poor nutrition—does the pathogen get out of hand. Some bacteria are apparently harmless, and some are actually necessary.

Your job, then, is to maintain the natural state of your goat's health by providing her with the proper feed and environment.

This isn't meant to imply that goats never get sick, that if they do it's because we did something wrong, or that there's nothing we can do for them. While you might go for years without seeing any health problems, if you have a large number of goats (or live with a few of them long enough), you're almost certain to encounter some ailments. However, you don't need a medical degree to raise goats: if an animal gets sick, all you need is the phone number of a veterinarian.

One of the most common complaints I, as an editor, hear from goat people, is "My vet doesn't know anything about goats." In some cases, the vet just doesn't care about goats. In any event, anyone who has graduated from veterinary school knows more about animal diseases than the rest of us do. I believe in making use of their knowledge…and if you have a good working relationship with a capable veterinarian, he or she will be glad to share much of that knowledge with you.

Here is a brief discussion of some of the conditions you're most likely to encounter.

ABORTION

Research has indicated that Bang's disease *(Brucella abortus)* is extremely rare among goats in the U.S., but Bang's disease tests are commonly required for showing, shipping, and for selling milk. Goats do abort, however.

If abortions occur early in pregnancy, the cause is apt to be liver fluke or coccidia. Liver flukes are a problem only in isolated areas such as the Northwest, where wet conditions favor them. Coccidia can be transferred by chickens and rabbits, both of which should be kept away from goat feed and mangers. (See *coccidiosis.*)

Abortion is more common in late pregnancy. The cause can be mechanical, such as the pregnant doe being butted by another or running into an obstruction such as a manger or a narrow doorway.

Certain types of medication can cause abortion, including worm medicines and hormones (such as are contained in certain antibiotics). Medicate pregnant animals with caution.

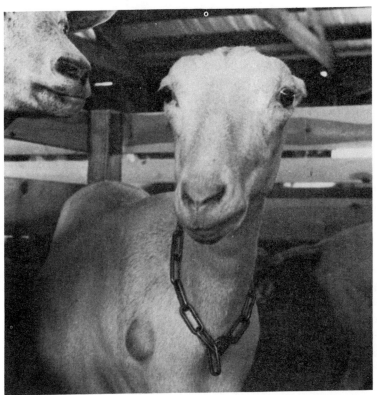

This La Mancha has an abscess on her shoulder. This is probably the most common goat ailment.

ABSCESS (*Caseous lymphadenitis*)

An abscess is a lump or boil, often in the neck or shoulder region, which grows in size until it bursts and a thick pus is exuded. In goats this is usually caused by *corynebacterium ovis* (also called *corynebacterium pseudotuberculosis*). It's common in sheep, and is the major disease of dairy goats in the United States.

With sheep, it's believed that the bacterium enters the body mostly through skin abrasions, principally caused by shearing, but unhealed navels, and docking and castration wounds are also suspect. Goats most likely pick it up from eating contaminated feeds, although external parasites that break the skin have also been suspected. The superficial lymph nodes are usually the primary sites of the abscesses. That's why we generally see them under the ears, on the throat, in front of the shoulders, and on the flank. However, the visceral lymph nodes can also be affected, and of course we can't see those without opening up the animal.

By the time we can see an abscess as a small lump, the animal might have been infected for several months. In some cases the lump will become as large as a grapefruit. What's happening is that the affected lymph nodes become enlarged by abscesses containing a cheesy green odorless pus. In older lesions this pus becomes a dry, firm capsule.

TREATMENT

Eventually the abscess will rupture by itself—causing further contamination and the spread of the highly contagious disease. Before it bursts, isolate the animal, make an incision—preferably a vertical one and as low as possible to promote drainage—with a sharp, sterilized knife, and carefully squeeze out the pus without getting any on your skin. Collect it in a paper towel and burn it immediately. Flush the wound with antiseptic solution and keep the affected animal isolated.

If the abscess is in the throat behind the jaw or under the ear, a veterinarian should perform the procedure, since these areas contain major blood vessels and nerves.

Dr. Sam Guss recommended four daily IM injections of 100,000 units of penicillin with dihydrostreptomycin after treating the abscess.

With any luck—and good management and strict sanitation—you can control the problem. However, some herds never eradicate *caseous lymphadenitis*. Then it might be necessary to raise kids completely separated from older animals until you can build up a diesase-free herd—or to depopulate, thoroughly disinfect the premises and start over with animals known to be free of the problem.

BANG'S DISEASE *(Brucellosis)*

Brucellosis is a contagious disease primarily affecting cattle, swine, sheep, goats, and dogs, and characterised chiefly by abortion. The first *Brucella* infection to be recognized was caprine brucellosis, or Malta fever, in 1887.

While it's prevalent in many countries where goats are common, brucellosis is extremely rare in the United States, but it's still considered a dreaded disease. Brucellosis immunizations are considered one of the standard routine health care requirements.

Abortion occurs in about the fourth month of pregnancy. Diagnosis requires a bacteriologic examination of the milk or the aborted fetus, or a serum agglutination test. There is no cure: the disease is treated by the slaughter of reacting animals.

Many people are concerned about brucellosis, but several years ago *Countryside* magazine investigated the thirteen cases reported in USDA annual statistics. These were in three herds, in Arizona, Indiana, and Ohio. A check with officials in Arizona and Indiana showed that their cases were in fact clerical errors. The one goat in Ohio was classed as positive on the test, suspicious on the retest, and after being slaughtered and subjected to a tissue test, negative.

Goats that have come up suspect after a Bang's test have invariably been pregnant or recently freshened. Subsequent tests are negative.

Of the two million goats slaughtered under Federal Meat Inspection from 1960-65 and 1969-71, not one case was found.

Nevertheless, to be absolutely certain about the safety of the milk from your goats, you can have them tested for Bang's, and you can pasteurize the milk.

BLOAT

Bloat is an excessive accumulation of gas in the rumen and reticulum resulting in distension. If you've just turned the goat out on a lush spring pasture or if she found out how to unlock the door to the feed room, anticipate bloat.

As always, the best cure is prevention. Feed dry hay before letting animals fill up on high-moisture grasses and clovers. Don't feed great quantities of succulents such as green corn stalks if the animals aren't used to them.

Bloat is caused by gas trapped in numerous tiny bubbles, making it impossible to burp. A cup of oil—corn, peanut, or mineral—will usually relieve the condition. A handful of bicarbonate of soda will help. In extreme cases it may be necessary to relieve the gas by making an incision at the peak of the distended flank, midway between the last rib and the point of the hip, holding the wound open with a tube or straw.

CAE *(Caprine arthritis encephalitis)*

Virtually unknown before 1979, CAE has caused near panic in some goat circles since then.

In one common form of the disease, the first signs are usually minor swelling in the front knees. The swollen knees become progressively worse, and the animal just seems to "waste." The lungs may become congested, and eventually all the body's systems give up. CAE is an incurable contagious disease, and some goats that have the virus do not show symptoms but are still carriers.

You can have your goats tested for CAE. If they test positive, there is nothing you can do for them, but because the infection is spread in the neonatal period, you can build and protect a "clean" herd by following this regimen:

1. Be there when the kids are born. Deliver them onto clean bedding, preferably newspaper rather than straw. Don't break the amniotic sac before the kid is delivered: the fluid in the sac is not infected, and the sac prevents the kid from swallowing or inhaling infected cells.

2. Put each newborn kid into clean, separate boxes. As soon as you can, wash each one in warm, running water to eliminate the possibility of any infected fluid on the body being ingested. Keep them separate until they're clean and dry.

3. Within half an hour or so, feed the kids colostrum—from a goat you *know* is free of the virus, or from a cow; in case of emergency, feed a home-brewed colostrum replacer (see page 122).

4. Feed only pasteurized milk, cow milk, or sheep or goat milk replacer. (Milk replacer made for calves is not high enough in fat for sheep or goats.)

5. Keep the kids separated from other goats and practice strict hygiene: sanitize feeding utensils; care for the kids before handling or walking among the older animals; and wash your hands, change boots, etc., before going among the kids.

CASEOUS LYMPHADENITIS

(See Abscess)

COCCIDIOSIS

This disease is caused by microscopic protozoans (coccidia) found in the cells of the intestinal lining, and is therefore a parasitic disease (see Parasites, Internal). It usually occurs in kids one to four months old, and usually in crowded and unsanitary pens. The most common symptom is bloody diarrhea, although kids with coccidiosis are usually weak and unthrifty. Your vet might recommend a sulfonamide in the feed mixture to treat this condition.

CUTS

Cuts, punctures, gashes, and other wounds can almost always be avoided by good management. They can be caused by such hazards as barbed wire, horned goats, junk or sloppy housekeeping, and other conditions under the control of the goat caretaker. Still, accidents can happen.

Clean such wounds with hydrogen peroxide and treat with disinfectant such as iodine. Use your own judgement to decide if stitching is required, or get the animal to a vet.

ENTEROTOXEMIA

Enterotoxemia is also called pulpy kidney disease and overeating disease. An autopsy soon after death will often show soft spots on the kidney.

The usual symptom of enterotoxemia is a dead kid. There is always misery, and almost always a peculiarly evil-smelling diarrhea. With some strains there may be bloat or staggering.

It is caused by a bacterium that is always present, but which, when deprived of oxygen in the digestive system, produces poisons. The proper conditions can be induced by overfeeding. Goats build up resistance to the poisons produced in small, regular amounts, but they can't handle sudden surges of them.

There are six types of *Clostridium perfringens* bacteria that cause enterotoxemia. Types B, C, and D cause the most trouble, with type D most often affecting sheep and goats. Antitoxin can be administered if you get there fast enough, but death is usually swift. Where enterotoxemia is a problem, vaccines are available from your veterinarian. Annual booster shots are required, and the kids will get antibodies from their dams. The best prevention is proper feeding on a suitably bulky, fibrous diet.

GOAT POX

The symptom is pimples that turn to watery blisters, then to sticky and encrusted scabs on the udder or other hairless areas such as the lips. It varies in severity.

Pox can be controlled by proper management, especially involving sanitation. Infected milkers should be isolated and milked last to avoid spreading the malady to others. Time and gentle milking are the best cures.

Very similar conditions can be caused by irritation. I have seen cases caused by dirty, urine-soaked bedding and by the use of udder-washing solutions that were too strong. In these cases the cure is wrought by removing the cause. An antibiotic salve will keep the skin supple and prevent secondary infections. Traditional treatment is methyl violet to dry up the blisters, but this drying can make the udder painful.

JOHNNE'S DISEASE

This, like CAE, is one of the "wasting" diseases. Frequently the only symptom is skinniness or loss of condition. Scouring is a typical symptom in cattle but not always in goats. It cannot be diagnosed accurately in goats except by autopsy. The intradermal Johnin test used on cows has come up negative on goats that actually had the disease as determined by autopsy.

The disease apparently infects young animals, either by interuterine transmission, congenitally at birth, or by mouth. If the disease follows the pattern it does in cows, adult animals can be sources of infection even if they do not show clinical signs of disease. The kid infected at birth typically won't start getting sickly for one and a half to two years.

There is no reliable test, and no cure. Prevention consists of starting with a clean herd and keeping it that way.

KETOSIS

Ketosis includes pregnancy disease, acetonemia, twin lambing disease, and others. Symptoms include a lack of appetite and listlessness. Ketosis occurs in the last month of pregnancy, or within a month after kidding. Its primary cause is poor nutrition in late pregnancy—but it's most likely to affect fat does, especially those that get little exercise. A dairy goat should never be fat, and nutrition is particularly important when the unborn kids are growing rapidly and making huge demands on the doe's body.

Treatment consists of administering 6-8 ounces of propylene glycol. This may be given orally twice a day, but not for more than two days.

In an emergency situation, try a tablespoonful of bicarbonate of soda in 4 ounces of water, followed immediately by one cup of honey or molasses.

Once the advanced stages have developed, no treatment is effective.

LICE

Suspect lice if your goat is abnormally fidgety and has a dull, scruffy coat. Fresh air and rain are good preventatives. Ask your vet for a louse powder approved for use on dairy animals. For an old-time cure, apply two parts lard to one part kerosene.

MANGE

Mange is indicated by flaky, scurfy "dandruff" on the skin. It's accompanied by irritation. Hairlessness develops and the skin becomes thick, hard, and corrugated. The condition is caused by a very tiny mite. There are several types of mange. Demodectic is probably most common, and can be stubborn. Nicotine, arsenic sulfur and creosote dips have been used in the past, but a solution of 0.06% lindane is most common. These agents can be dabbed on.

Scurfy skin can also be the result of malnutrition or internal parasite infestation.

MASTITIS

Symptoms: a hot, hard, tender udder; milk may be stringy or bloody. (Routine use of a strip cup before milking will alert you to abnormal milk.) Mastitis may be subclinical, acute, or chronic. Hard udders (usually just after kidding) that test negative for mastitis are referred to as congested, and usually disappear. Congested udders are best cleared up by letting the kids nurse and massaging the teats and udder for three to four days after parturition. In mastitis the alveola or milk ducts are actually destroyed. Since it's necessary to identify the bacteria involved, the services of a vet are required.

Mastitis can be caused by injury to the udder, poor milking practices, or transference by the milker from one animal to another. Teat dips have proved of great value in controlling the disease among cattle, although the solutions must be diluted for goats.

Home tests for mastitis are available from veterinary supply houses.

MILK FEVER *(Parturient Paresis)*

Symptoms: anxiety, uncontrolled movements, staggering, collapse, and death. Usually this occurs within forty-eight hours of kidding. It's caused by a drastic drop in blood calcium, which is related to the calcium level of feed consumed during the dry period, and even to incorrect feeding of young animals. It can be brought on by sudden changes in feed or short periods of fasting. Curing milk fever requires quick action and a vet, who will administer calcium borogluconate intravenously.

PARASITES, EXTERNAL

(See also Lice, Mange)

External parasites are generally much less of a problem with goats than are internal parasites.

Lice are almost universal, but mild infestations cause little harm to well-nourished animals. A badly infested goat will rub against posts and other objects, have dry skin and dandruff, and can lose a great deal of hair. Lice can be controlled by dusting, spraying, or in large herds, dipping. All members of a herd must be treated at the same time to control lice. Use a dust approved for dairy animals. With plenty of sunlight and fresh air, lice seldom become a serious burden.

Screwworms can be a problem, particularly in the South and Southwest. They survive on living flesh, and normally depend on wounds—including dehorning and castration—in order to enter an animal. Castrating with the Burdizzo and timing dehorning so the wounds will be healed before fly season will help prevent screwworm infestations.

Goats can be affected by mites, which produce the diseases mange and scabies. Sarcoptic mites, responsible for sarcoptic mange, can affect all species of animals: demodectic mange and psoroptic ear mange are specific to goats. Scabies is very rare in goats. It is not infectious to other animals or humans.

PARASITES, INTERNAL

"Worms" of various kinds are perhaps the most widespread and serious threat to goats' well-being, but only when they're present in large numbers. Many goat raisers worm goats regularly with a favorite anthelmintic, but alas, it isn't quite this simple.

The list of internal parasites that infest goats is quite long. It includes bladder worms, brown stomach worms, coccidia, four species of cooperias, hookworms, liver flukes, lungworms, nodular worms, stomach worms, tapeworms, whipworms, and others. Some are quite common in certain areas and rare elsewhere.

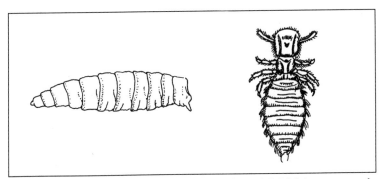

The screwworm larvae (left) and the common sheep tick (right) are two pesky external parasites.

Not all of these are affected to the same degree by a specific wormer or anthelmintic. This means a fecal test is required so your vet—or you, if you have a microscope and an interest in such matters—can determine which parasites are present, and therefore which veterinary product to use.

In addition, many parasites can build up resistance to a given anthelmintic: it's usually necessary to periodically rotate the products used. Some of the more common vermifuges include Tramisol® (levamisol), TBZ® and Omnizole® (thiabendazole), Camvet® (cambendazole), Panacur® (fendbendazole), Telmin® (mebendazole), and Benzelmin® (oxfendazole). As if matters weren't complicated enough already, some cross-resistance to some of these drugs in the same chemical family has been reported.

The best method of dealing with parasite problems or potential problems consists of two simple steps: 1. regular and close inspection of your goats; and 2. periodic fecal exams.

INSPECTING GOATS

Pay attention to your goats' mucous membranes—particularly the eyes—and the gums. The "whites" of the eyes should not be white, but pinkish or red. If they're white, or if the gums are pale pink or grey, this indicates that the goat is anemic. The likely cause is worms.

This isn't foolproof either, though. *Muellerius,* for example, doesn't cause anemia: it destroys lung tissue. (And *muellerius* is not affected by TBZ or Tramisole, two of the most common goat wormers. If you simply worm with these on a regular basis, you might enjoy a false sense of security.)

TREATMENT

Anthelmintics come in various forms: boluses (large pills), pastes, gels, powders, crumbles, and liquids.

Boluses are popular, but many goats refuse to take them, and because they can choke a goat, some people refuse to use them. They can be administered with a balling gun, or try hiding a bolus inside a gob of peanut butter.

Drenching, or administering a medication from a bottle, can also be risky, but it's almost a required goatkeeping skill. It's important to: 1. give a little at a time and allow the goat to swallow in between; 2. give in the left-hand corner of the mouth; and 3. never raise the head—keep the muzzle level.

When administering paste wormers, be sure the goat doesn't have anything in its mouth, including a cud. Put the paste in the back corner of the mouth on the left side. If the goat wants to shake her head and fling the paste out of her mouth, hold her muzzle gently and massage her throat until you're sure the wormer has been swallowed.

FECAL EXAMS

The best advice that can be given to beginners, or to anyone who doesn't want to become an expert on worms, is to have laboratory fecal exams performed twice a year, and to follow the advice of a veterinarian. A schedule can be set up based on the life cycles of the specific parasites present and the anthelmintics chosen.

One sample deworming schedule that has been used in a large herd in the Southeast can serve as an example, but the owner of this herd emphasizes that it's merely an outline: you should set up your own schedule based on examinations of your own goats, the results of fecal tests, and the advice of your local vet.

1. One week before kidding: deworm with TBZ.
2. Two weeks later: deworm does with Panacur or Tramisol.
3. One to two months later: deworm does with Panacur, and kids with Camvet. Give the kids a second dose two weeks later.
4. Midsummer: have fecal tests run and if necessary, use the drug applicable to the specific parasite problem present.
5. Four weeks before breeding: deworm with Tramisol.
6. Two weeks later: deworm with Panacur or Benzelmin. Never deworm a goat during the first half of pregnancy.

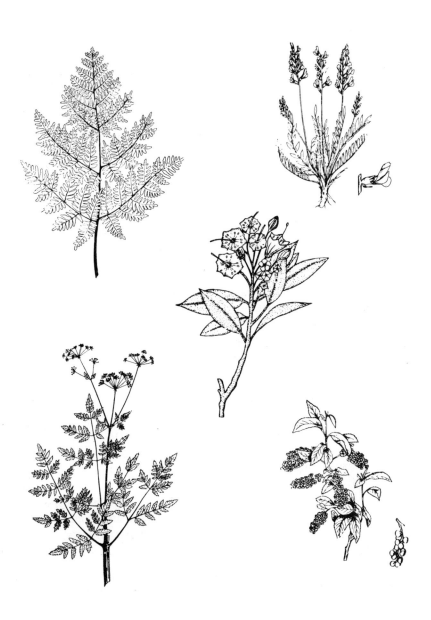

Five poisonous plants. Upper left: *break fern.* Upper right: *locoweed.* Center: *mountain laurel.* Lower left: *European hemlock.* Lower right: *wild cherry.*

POISONING

Symptoms are vomiting, frothing, and staggering or convulsions. Because of the nature of a goat's eating habits, poisoning from plants is rare: she takes a bite of this and a taste of that and will seldom eat enough of one poisonous plant to cause much damage. To learn what plants in your particular locale are poisonous, check with your local county agent. Some to watch out for are bracken, locoweed, milkweed, wilted wild cherry leaves, and mountain laurel. Rhubarb leaves are poisonous.

Lead poisoning is a possibility because goats are likely to chew on wooden surfaces. Use whitewash or lead-free paint (although that too can be dangerous if enough is ingested), and don't gather goat browse from roadsides where concentrations of lead from automobile exhausts are heavy. Avoid taking feed of any kind from along roadsides that might have been sprayed for weed control. Don't feed Christmas trees to goats because many of them are sprayed with toxic substances. If your neighbor sprays any crop, keep your goats away from any area that might have been contaminated by drifting spray. Many seeds are treated and poisonous. Every so often we hear of fertilizers or insecticides or other chemicals that look like feed additives killing off whole herds of cows when someone mistakenly grabs the wrong bag. Be careful.

Antidotes depend on the poison. Call a vet.

SCOURS (DIARRHEA)

Scours in newborn kids can indicate any of a number of problems, including failure to ingest colostrum soon after birth, lack of sanitation, inadequate nutrition of the doe during gestation, feeding excessive amounts of milk, and feeding low-quality milk replacers. The mortality rate is high, and swift action is required.

Since death results from dehydration and shock, the first goal is to restore electrolyte balance. Electrolyte formulations are available from drug companies, but in an emergency a suitable solution can be mixed from ingredients found in any kitchen.

1 gallon warm water
2 teaspoons table salt
1 teaspoon baking soda

8 tablespoons honey, white corn syrup, or crystalline dextrose (Never use cane sugar!)

Neomycin, nitrofurazone, or chloramphenicol can be added, or given separately.

If the kid is too weak to nurse, this can be administered with a syringe or stomach tube. Give 1 to 2 cups per 10 pounds of body weight per day until the scours clears up. Don't feed milk during this period.

The above solution is called hypotonic: it contains electrolytes in roughly half the concentration of electrolytes in the blood. This is given only by mouth. However, a veterinarian can administer isotonic electrolyte solutions in which the concentration is the same as in the blood, intravenously.

Another home remedy provided by a vet:

1 cup buttermilk	1 teaspoon cocoa
1 raw egg	1/4 teaspoon baking soda

Mix in a blender or shake well in a jar. Bottle feed one-fourth of this every two to three hours. One crushed bolus of neomycin can be added to this mixture.

TETANUS

Goats with tetanus will have their heads held up in an anxious posture and will be generally tense. There is difficulty in swallowing liquids, and muscular spasm. Death occurs within nine days. Tetanus or lockjaw requires a wound for the germ to enter, but it can be something so simple the caretaker doesn't even notice it. Treat all punctures and cuts with iodine, and pay special attention to the navels of newborn kids. If you have horses, or if horses or mules have been on your farm in the past twenty years, tetanus vaccination is recommended. Treating tetanus is a job for a vet.

WHITE MUSCLE DISEASE

This is caused by a lack of selenium. It most commonly affects healthy, fast-growing kids less than two months old, although problems can occur in mature animals. Stiff hind legs can indicate white muscle disease in kids, but it can also cause sudden death in animals less than two weeks old.

Soils in many parts of the United States are deficient in selenium. Providing this mineral (5.0 milligrams, oral or intramuscular) two to four weeks before kidding will prevent deficiencies in the doe and in the kid for its first month. Kids should receive injections at two to four weeks.

There is a relationship between selenium and vitamin E. The two are usually administered together.,

Note that in excessive amounts, selenium is a poison. In some areas soil levels are high enough so that plants grown on them cause toxicity resulting in paralysis, blindness, and even death.

WORMS

(See Parasites, Internal)

To repeat, if you start with healthy goats and give them proper care, chances are good that they'll have few health problems.

If you do have a sick goat, don't hesitate to call a veterinarian. It's easy for amateurs to misdiagnose animal illnesses. Furthermore, if you try to get a veterinary education from one or two books, try all the home remedies you can find and only call the vet when the animal has three feet in the grave, don't expect a doctor to perform miracles. Vets carry drugs, not Bibles.

CHAPTER 9

The Buck

Goats won't produce milk without kidding and they won't kid without being bred. That requires the services of a buck, and that entails a whole 'nuther look at goat raising.

You'll need a buck, but that doesn't mean you'll have to buy one. In fact, beginners are usually advised to forget about keeping a buck. There are many reasons.

One practical reason is expense. A buck requires the same amount of housing, feed, bedding, and grooming as a doe. Therefore if you have one doe and one buck, the cost of your milk is double what it would be without the buck. The number of does you need to justify the expense of having your own herd sire depends pretty much on your particular situation. If you live in a remote place where there is absolutely no stud service available within a "reasonable" distance (and this might be a few miles or a few hundred miles, again depending on you), it might be necessary to have a buck even for just a few does. Expensive, but more economical than the alternatives.

Even then, you'll face problems. Good bucks are expensive, and serious breeders won't sell any other kind. They're worth it, of course, because the buck you use this year will affect your herd for years to come. While a doe can be expected to produce one or two kids a year, if you only use one buck, his genes will be passed on to every kid you raise. The buck is truly "half the herd."

IMPROVING THE BREED

A good breeder won't sell you a "cheap" buck: if it isn't good enough to improve most goats of its breed, it is slaughtered at birth or castrated.

101

A Toggenburg buck.

A Nubian buck.

Unfortunately, there are too many people raising goats who are not good breeders.

This deserves some explanation, because there are two approaches to raising livestock, and not only are most beginners somewhat puzzled by the differences, but their natural inclinations frequently attract them to the less desirable approach.

We're speaking here of breed improvement. While most emphasis on breed improvement naturally comes from people who are involved in showing their animals, be they rabbits, dogs, cows, or goats, there is more than ample evidence to prove that the "commercial" or homestead or backyard producer has every bit as much to gain from striving for improvement...and perhaps more. And there is really very little—or nothing—to lose.

I have found it frustrating to deal with people who place little or no emphasis on breed improvement, or who even actively belittle the fancy-pants show enthusiasts as if their interests were somehow contradictory. Nothing could be further from the truth.

For proof we need only turn to the commercial dairy (cow) farmer. Almost invariably these practical, tough-minded, cost-conscious farmers use the best purebred registered bulls they can find. They may not have the slightest intention of ever showing a cow or of raising purebred cattle (although some of them are finding that purebred cows are valuable for the same reasons purebred bulls are practical). They use purebreds because it pays off in the milk pail.

Milk production per cow has just about doubled since the last century. While some of this is due to feeding practices and other management details, a large share of the credit must go to genetics.

Similar progress has not taken place in goats. Maybe it's not even a desirable goal. But no lover of goats can deny that there are entirely too many half-pint milkers around (and being sold to unsuspecting novices who have heard that goats give a gallon of milk a day). The reason is simply poor culling practice...often starting with the selection of the buck.

Just as a good buck will improve many future milking does, a poor buck will drag them down. Cull a poor buck today, and save yourself the trouble of culling dozens or hundreds of poor does in the future.

Now, it's true that showing, and in fact the entire registration system, sometimes aids and abets this: there *are* goats that place high in the show ring but don't produce enough milk to pay for their grain ration. It is true that too many goats are registered (and sold for high prices and allowed to

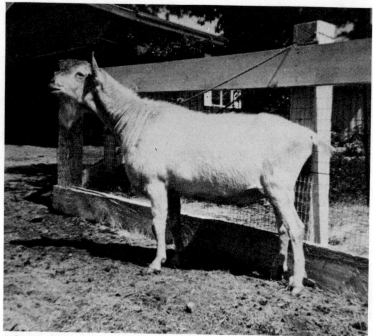

A Saanen buck.

produce more high-priced goats) simply because they are purebred. If you just want milk, you can ignore the faults in the system, but don't throw out the baby with the bath water. Ignore the faults, but not the entire system, if you want to produce milk efficiently.

CULLING POOR BUCKS

That means following the system as it was meant to be followed. Breed the best to the best and cull the rest. Cull means destroy: it doesn't mean passing on substandard animals to the first sucker who comes along, or selling them as pets only to have them (or their offspring) find their way back into someone's dairy. Too many people take the short-term view of the economic loss incurred, and as a result short-change themselves, and future goat raisers, for the long term.

Your chances of improving your herd are practically nil if you breed your does to a neighbor's nondescript pet buck simply because he happens to be cheap and available, or if you buy a buck just because he's cheap. You aren't going to milk the buck, but never forget that you're going to milk his daughters, and hopefully many future generations. If he doesn't have the genetic potential for milk production, his daughters won't have it either. If the buck is not much better than the doe, you're not working for breed improvement. In fact, you're not even breeding goats: you're merely using the buck as a doe freshener to get a little milk.

Actually, there have been cases where goatkeepers (as distinguished from goat *breeders*) who are only interested in freshening does because they only want milk don't even keep replacement animals. Since the buck has no effect on the lactations of the does he breeds, he would have no effect on the herd if all his offspring were routinely butchered. This seldom happens, however, and the downgraded goats are foisted upon a world that already has too low an opinion of these valuable animals.

PEDIGREE

So how do you choose a buck that will produce superior offspring? You start by examining his pedigree, the record of his ancestors. If it's milk you want, make sure there's milk in that pedigree.

While there admittedly are lovely grades that make milk by the ton, there is no way of knowing who their ancestors were or how good they were. A pedigree and milk production records of several generations of forebears might not be insurance, but they're valuable management tools, and much better than flying blind.

INBREEDING

Now, let's suppose you purchase a fine buck of impeccable breeding, excellent health, and ideal conformation. Are your problems over? Not quite.

If you have four does, a fair average for a homestead herd, you can expect four doe kids the first year. Chances are you'll want to keep one or more of them: after all, didn't you buy the high-powered buck to improve your herd?

But then, it's evident that the next breeding season you'll be making father-daughter matings. This isn't necessarily bad. While many people attach the human-oriented stigma of incest to such matings, in the hands of an expert they are actually the surest and fastest way to breed improvement. But few beginning or backyard goat raisers are experts in animal genetics, if only because a herd of a few animals would require years and years to give the breeder the experience necessary to be an "expert." It's sufficient to say here that inbreeding emphasizes faults as well as good points; it's nothing to be dealt with haphazardly. (Actually, there is some evidence to suggest that inbreeding affects goats less than some other animals. But are your original goats good enough to be perpetuated—or should they be upgraded by outcrossing?)

So when your herd sire's first daughters come into heat you'll want another buck.

MINIMIZING FAULTS

This is a good place to note that no animal is "perfect." All have faults of one kind or another, to a greater or lesser degree. It's the job of the breeder to eliminate those faults as much as possible in future generations, while at the same time preventing new ones from showing up.

An illustration of this would be a doe with very good milk production, but a pendulous udder. That udder fault is going to shorten her productive life, it will make her more liable to encounter udder injury and mastitis, and so forth. So you'll want to breed such a doe to a buck that tends to throw daughters with extra-nice udders, in hopes that the offspring will have both good production and acceptable udders. They won't be extra-nice in view of the genes contributed by the dam, but they can be improved by proper buck selection.

The problem here is that, with four different goats in a small backyard dairy, there are likely to be at least four different faults! It's unlikely that even a very good buck will be strong enough in four different areas to compensate for all of them. From the standpoint of breed improvement, then, each doe in your barn is likely to be best matched by a different buck.

These are real and important and practical considerations. But we must also mention some more commonly voiced objections to keeping bucks.

PROS AND CONS OF KEEPING BUCKS

Unlike does, bucks *do* smell, especially during breeding season. Girl goats (and some goat people) are inclined to like the aroma, but it will not only hang over your barnyard: it will pervade your clothes, and even your living room furniture will get to smelling like ripe billy goat, which for most people is less than desirable.

Also of interest to people who are new to goats are what they often call the buck's "objectionable, disgusting habits." Most city people are shocked when they find out that the cute and playful buck kid grows—astoundingly rapidly—into a male beast who not only tongues urine streams from females (and makes funny faces afterwards), but who also sprays his own beard and forelegs with his own urine. This is natural goat behavior, but be that as it may, even many broad-minded people find it difficult to accept gracefully. Needless to say, the loveable buck kid loses a few friends when he reaches this stage.

Bucks are powerful animals—I've seen them snap 2 x 6's just for kicks—and one that has not been raised properly or finds himself in an untenable position *can* be a dangerous animal. (I have never owned a buck that was any more hostile or aggressive toward humans than a doe— although they haven't been effeminate either, which would be a fault in a buck. But enough other people speak of "mean" or "dangerous" bucks, so it seems likely that they exist, and you should be forewarned.)

Because they are powerful, and because of their natural sexual instincts, a buck requires much more elaborate and expensive housing than the does, especially during the breeding season. They must be housed separately if only to avoid off-flavored milk. And an inadequately penned buck will soon be found with the girls.

In spite of all this, there are many practical and logical reasons for keeping bucks even for small herds, and many people do.

While there are many advantages to buying a proven sire—a buck you know is not sterile and who throws daughters with the traits you want in your herd—such bucks are either very expensive or old and otherwise worn out. Most bucks are sold as kids, fresh off their dam's colostrum.

As mentioned, most good breeders dispose of buck kids at birth, even very good ones, because there is little demand for them: half of all kids born are bucks, and only a small fraction of those are needed for breeding. Only the very best are kept, and almost invariably these have been reserved far in advance.

Buck kids are raised very much like doe kids. They grow a little faster, but take longer to fully mature. However, even though a buck may not stop growing until he's three years old, he is capable of breeding a doe by the time he's three or four months old. Don't let his size or his kid-like demeanor fool you! This means separate penning is necessary almost from the beginning, and at least by two months of age.

A buck can be used for limited service even before he's a year old. Most authorities say he should be limited to ten to twelve does his first year. A mature buck can service more than a hundred does a year, according to some reports, and of course, with artificial insemination (AI) many more than that. Let's discuss this next.

CHAPTER 10

Breeding

A goat obviously must be bred in order to produce kids—and milk. Those 155 or so days between breeding and kidding are extremely important to the goat, and for the first-time goat owner especially, they are anxious ones.

THE DOE'S CYCLE

It all starts with the doe's estrous cycle, or heat periods. Goats (and most other animals) "cycle"; that is, they are fertile only for relatively short periods at more or less regular intervals. Unlike cows or hogs, which come into heat year around, goats generally come into heat only in the fall and early winter. A doe will accept service from a buck only when she is in standing heat, usually. If she is not in heat, copulation won't result in pregnancy anyway because the sperm and the ova aren't in the right place at the right time.

Seasonal breeding has decided advantages for animals such as deer and goats. Their young are born when it isn't too cold; there is plenty of lush, milk-producing feed for the mother, and tender grasses and leaves for the young to be weaned on; and the offspring are fairly strong and independent by the time the weather turns harsh again. Desirable as such an arrangement may be for wild animals, it puts the dairy goat farmer in a bind.

THE LACTATION CURVE

In Chapter 2 we examined the lactation curve. If you plot such curves for several goats, all of which have been bred at more or less the same time, it's apparent that the goat farmer will be swimming in milk during part of the year and dry as a bone in another part. This is perhaps the single most serious drawback to commercial goat dairying. If people want to buy goat milk, they want to buy it regularly, not just when it's in season.

The normal lactation curve is reinforced by seasonal curves that are equally normal in both cows and goats, due to feeding conditions, weather, and other factors. Animals simply produce more milk in summer than in winter.

Add to that the fact that more people *want* goat milk in winter than in summer, and it's easy to see that the poor goat farmer has a problem.

Backyard dairy operators and homesteaders share the same dilemma, to some degree. If you have just one goat, even if she has a lactation of ten months, you'll be without any milk at all for two months of the year. With two goats you can attempt to breed one in September and one in December. Then, theoretically, you will never be without milk, but a look at the lactation curves plotted together will show that your milk supply will be far from steady. You'll have too little or too much far more often than you'll have just enough.

The point here is that you'll want to have your does bred as far apart as possible, but while still avoiding the risk of having a doe miss being bred at all. With some does, in some years, even December may be pushing it: they simply won't come in heat again until September. There *are* kids born in every month of the year, but as a practical matter for the small-scale raiser, you can't count on out-of-season breedings.

("Milking through," or having a goat that milks for a year or more without drying off, is another possibility. Such animals do exist, but they're far from common.)

DETECTING HEAT PERIOD

For many beginners, and especially those with only one or two goats, it's very difficult to tell when a goat is in heat. The usual signs are increased tail-wagging, nervous bleating, a slightly swollen vulva sometimes accompanied by a discharge, riding other goats or being ridden by them, and sometimes by lack of appetite and a drop in milk production. If a buck is nearby there will be no doubt: she'll moon around the buck pen side of her yard like a lovesick teenager.

If you lack a buck and have trouble detecting heat periods, or just want to make very sure she's in heat before you make a several hour trip to a buck, you might use this trick: Rub down an aromatic buck with a cloth, or tie one around his neck for a couple of hours. Gingerly poke it into a canning jar and screw the lid on. When you suspect your doe is eager for male companionship, give her a whiff of that cloth and your suspicions will quickly be confirmed or denied.

But you can't breed a doe with canned buck aroma. If you don't have a buck you'll have to take her to one.

TRANSPORTING THE DOE

In the first edition of this book I mentioned that when we had neither a buck nor a pickup truck, we used to transport does in the trunk of the car. Some readers were aghast at this dangerous practice. Since it always worked fine for us, since there's no way I can afford a livestock trailer, and since a pickup truck (even with side racks) is far more hazardous for the goat, I still favor the car trunk. But you decide for yourself.

If your car isn't too fancy, or if you really love goats, she can ride in the back seat. If she's lying down, she will not "disgrace herself," as one puritanical old goat book put it. Some goats tend to get carsick standing up and will be too woozy to be interested in breeding after the trip. On the other hand, some enjoy riding in a car as much as dogs do, even to the point of sticking their heads out the window...which could easily cause an accident on a busy freeway among drivers who have never seen a goat riding in a car.

ARTIFICIAL INSEMINATION (AI)

There is another possibility that interests many people: AI, or artificial insemination. This is a relatively new development in the goat world, it's not as common as artificially inseminating cows, and it's not 100 percent foolproof, but it has great potential.

From the standpoint of breed improvement, there is nothing better. Anyone, anywhere, can use the finest bucks available, and at low cost. In many cases one straw of semen (a straw is the glass tube the semen is stored in) costs only a tenth of the same buck's standing stud fee, and you can use bucks that are thousands of miles away. Does can even be bred to bucks that are long dead: semen can be stored for years.

Inbreeding problems mentioned in the chapter on bucks are easily eliminated with AI.

Even goatkeepers who aren't overly concerned about inbreeding can readily see the advantages of not having to keep a buck or having to traipse all over the countryside with lovesick does in the car.

At the same time, AI isn't the final answer to every goat owner's breeding concerns. In many cases it will be necessary for you to do the inseminating yourself...and you won't learn how to do it by reading a book. If you'll have to buy (and maintain) a liquid nitrogen semen storage tank, your breeding costs will go up appreciably.

All of this is far beyond the basics of raising goats, and the scope of this book. It's a new and growing field, and changes are taking place rapidly. For up-to-date information, look to the goat periodicals for ads of artificial insemination companies and contact them. Of course, you can also get the information you need to get started from any goat raisers in your area who are using AI.

BREEDING PROBLEMS

A doe will be in standing heat for twenty-four hours, although this varies widely. If she is not bred, she will come into heat again in twenty-one days, although this too is an average that varies considerably.

If a doe is serviced and still comes back in heat, there could be several causes. She might not have been bred at the most opportune time. Maybe one more try will turn the trick. (It isn't necessary to leave the buck and doe together for long periods: if the doe is really in standing heat, one service is sufficient, and that won't take more than a minute...which sometimes seems silly after you've spent an hour on the road and still have to drive home again!)

Sterile bucks are rare, and if a buck is sterile, obviously none of the does he serves will conceive. However, sperm can have reduced viability at certain times due to overuse of the buck and other factors.

If a doe simply will not get bred, the most common cause is cystic ovaries—a growth preventing ovulation—and she is worthless. Overly fat does are often difficult breeders because of a buildup of fat around the ovaries.

Another serious condition, although it's not as common as we once thought it was, is hermaphroditism, or bisexualism. The goat looks like a doe externally, but it actually has male organs internally. Not all "hermies" have obvious external abnormalities, but carefully examine the vulvas of newborn kids. A growth about the size of a pea at the bottom of the vagina is abnormal. Unusual behavior in a normal-appearing doe kid is cause for suspicion: hermies are often overly aggressive or unusually withdrawn.

The word *hermaphrodite* goes back to Greek mythology and the story of the son of Hermes and Aphrodite, who became united with the body of a nymph, Salmacis, while bathing in her fountain. In goats, the condition is often related to the mating of two naturally hornless animals. Not only do the genetics get a little complicated, but many practical goat breeders claim the geneticists are wrong anyway.

Basically you must determine whether a naturally hornless buck is homozygous or heterozygous; that is, whether or not it inherited a gene for horns from either of his parents.

If either the buck's sire or dam was horned, he's heterozygous. If neither parent was horned, you can't be sure what he is without seeing a number of his kids. If any of his kids have horns, the buck is heterozygous. If all the kids are hornless, even out of horned does, chances are the buck is homozygous.

Theoretically, there can be no homozygous does because they'd be hermaphrodites and couldn't have offspring. Both types of bucks will produce some hermaphrodites when bred to hornless does, according to theory, but the homozygous hornless bucks will produce more.

All of this is of great interest to geneticists and large goat breeders and people who take a keen interest in breeding... but the average backyarder or homesteader is better off to follow the lead of the major commercial goat farms and just avoid hornless-to-hornless matings.

WHEN TO BREED

Doelings are sexually mature as early as three or four months of age. In most cases spring kids that are well developed and healthy should be bred when they weigh about 80 pounds and are seven or eight months old, which means they'll kid at one year of age. Being bred too early will adversely affect their growth and milk production; being bred too late does not contribute to their health and welfare; it's expensive to keep dry yearlings; and records show that does that kid at one year of age produce more milk in a lifetime than those held over. Many people mistakenly hold back young does because "they look so small" or because seven months seems so young. With proper nutrition, they'll produce healthy kids and keep on growing themselves.

DRYING OFF

The older doe will be, or should be, still milking when she's bred. But advancing pregnancy will cause most does to dry off. Some people who really want milk will continue milking a naturally drying-off doe as long as she's giving a few squirts; others figure even a pound isn't worth their trouble, stop milking, and the doe dries up.

In any event it's a good practice to dry off a doe two months before her kids are due. Milk production makes great demands on a doe's body. So do growing unborn kids. The kids—and the health of the doe herself—are more important than the milk. In most cases simply quitting milking and reducing the grain ration will cause the animal to dry off naturally. In cases of extremely heavy or persistent milkers it may be necessary to milk her out at intervals. Milk once a day for a few days, then every other day, then stop. Reducing or eliminating grain will help an animal dry off.

FEEDING THE PREGANT DOE

No good dairy animal—cow or goat—can eat enough during lactation to support herself and her production. That's why she requires a rest to build up her body before peeling off her own reserves to fill your milk pail. It's been said that for each pound of increase in body condition during the dry period, a Holstein cow will produce an extra 25 pounds of milk, a Guernsey 20 pounds, and a Jersey 15. We could anticipate proportional results with goats.

This is not to suggest that a pregnant animal should be overconditioned or fat. A dairy animal should never be fat: just in good condition. In fact, fat causes problems in pregnancy. But the goat needs a well-balanced diet. Too much feed produces kids that are too large to be easily delivered. Excess minerals in the doe's diet produces kids with too-solid bones, which also causes difficulty. A fibrous diet with rather low protein is ideal for the first three months of pregnancy when the kids are developing slowly.

Most of the kids' growth comes in the last eight weeks of pregnancy. During this period the ration should be changed gradually, not only because two-thirds of the kids' growth is taking place, but because this is when the doe needs to build up her own reserves for her next lactation. High protein still isn't required—about 12 percent will do—but there is definite need for minerals and vitamins, especially iodine, calcium, and vitamins A and D. Bulk such as is provided by beet pulp or bran is required; molasses will supply some iron as well as the sugar that will help prevent ketosis, and it will have a desirable slight laxative effect.

Finally everything is ready. The goat stork cometh.

CHAPTER 11

Kidding

The "miracle of birth" is aptly named. Like all miracles, it's invested with wonder, awe, excitement, and joy. There have been cases of people who would have nothing to do with goats—until they saw newborn kids frolicking in fresh clean straw and fell in love. (I'm married to one of them.)

There is little doubt that the first kidding season brings the new goatkeeper excitement that is hard to duplicate in today's plastic and artificial world. Most of them, judging from the mail I get and my own first experiences, are scared silly as parturition approaches.

Most of this fear comes, I believe, from reading books and articles describing all the things that can go wrong. You expect the worst. But goats have been having kids all by themselves for thousands of years. While problems are possible, 99 percent of all goat births are completely normal and won't even require your assistance. The chances for a normal birth are enhanced by proper feeding and management, especially during the latter stages of pregnancy.

ANTICIPATING THE DELIVERY

The average gestation period for goats is 145 to 155 days. Some experts say there is evidence that goats and sheep can control the time of birth to coincide with copacetic weather conditions. Other people say they control it, all right, but usually to have the kids and lambs arrive on the coldest, most miserable night of the year.

Start checking your animals frequently and carefully 140 days after breeding. When she is getting ready to kid the doe will become nervous, and will appear hollow in the flank and on either side of the tail. There may be a discharge of mucus, but this can appear several days before kidding. When a more opaque, yellow, gelatinous discharge begins, it's for real.

Kids can be felt on the right side of the doe. It's good practice to feel for them at least twice a day. As long as you can feel them they won't be born for at least twelve hours.

If you feel the doe regularly you'll be able to notice the tensing of the womb. After this, one of the kids is forced up into the neck of the womb, causing the bulge in the right side to subside somewhat. This will be noticeable only if you have paid close attention to the doe in the days and weeks before. The movement of the first kid also causes the slope of the rump to move into a more horizontal position. At this point you can expect the first kid within a couple of hours.

Many people look to "bagging up," or enlarging of the udder, as a sign of approaching parturition. This is unreliable. Some goats don't bag up until after kidding, and others will have a heavy milk flow far in advance. In some cases, if the udder becomes hard and tight, it might be necessary to milk out the animal even before kidding.

FACILITIES

Although humans try to take good care of their animals, we often complicate things. We've mentioned feeding of the pregnant goat, which can affect the ease with which she delivers. A free-ranging, experienced goat knows what to eat, but if we provide her food she must depend on our judgment. Likewise, the goat kidding outside on her native mountain range knows what to do when her time approaches. She is probably safer and in more hygenic surroundings on her mountain than in your barn. It's just about impossible to duplicate natural conditions for domestic animals.

There are innumerable instances of goat owners going to the barn for morning chores and finding a couple of dried-off, vigorous, and playful kids in the pen with their mother. But it's definitely preferable to have some idea that the kids are on the way and to make certain preparations for them.

The doe should have an individual stall for kidding. It should be as antiseptic as possible and well bedded with fresh, clean litter. Something softer than long straw is preferable if you have it.

Don't leave a water bucket in the kidding pen. It can be dangerous, and the mother isn't interested in drinking at this point anyway.

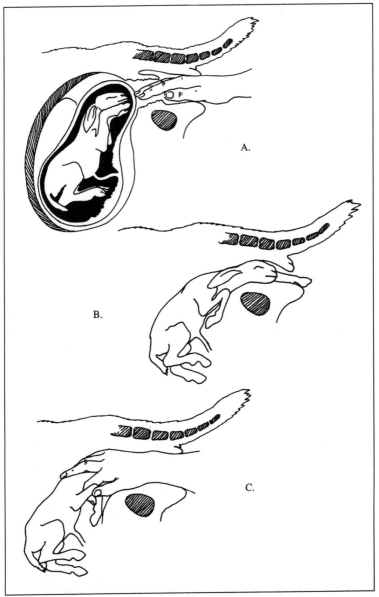

(a) A normal kid presentation. Kid below (b) has one foot left behind (middle) ; it can be pulled out by hand (c).

KID PRESENTATIONS

As mentioned, ninety-nine times out of a hundred there will be no problems and your assistance will be unnecessary and perhaps even unwanted. Normal labor can last anywhere from a matter of minutes to four hours or more, during which time you'll have nothing to do: you don't even have to boil water...unless you want to make a cup of tea.

In a normal birth the front feet and nose are presented first. Picture it in your mind's eye and you'll see that this presents a more or less cone shape which gradually distends the vagina.

Not all births are in this normal position. Sometimes the kid will try to be born with one foreleg bent back, effectively hooking it inside the womb; tail first; or in some multiple births, the kids and umbilical cords get all tangled up inside the womb.

Not quite "normal," yet not too unusual either, is a kid born backwards. This makes the delivery somewhat more difficult—that "wedge" is backwards—but the greatest danger is if the sac breaks before the kid is completely out. It could suffocate.

ASSISTING THE DELIVERY

If the doe has been struggling for a while, seems concerned, and nothing is happening, you'll have to help. Insert your disinfected, lubricated hand and arm into the birth canal to find out what's wrong. A germicidal soap will serve as disinfectant; mineral oil or K-Y jelly are lubricants.

If you've never seen a newborn kid this is not only scary, it's difficult to imagine what you're feeling for. Try to sort out the heads and legs, and if necessary, rearrange them in the proper presentation position. In most cases it will be a simple matter to "lead" the first one out the next time the doe strains. Pull, but very gently, working with the doe: otherwise hemorrhage might result. Chances are the others will come by themselves soon after.

A pessary, available from your veterinarian, should be inserted into the vagina after manual exploration to minimize the danger of infection.

Goats usually have two kids, but three, four, and even five aren't all that uncommon. If no more come within half an hour and the mother seems relaxed and comfortable, you can assume that's all there are.

If you happen to encounter a difficult birth—a dead kid that the doe can't expel, for instance—better get a vet or experienced neighbor to help. But again, such help is seldom needed. Be aware of what can go wrong, but please don't make yourself sick worrying about it beforehand.

In very cold climates, a box like this will keep chill drafts off newborn kids.

THE UMBILICAL CORD AND AFTERBIRTH

In most cases the umbilical cord breaks by itself, or the doe severs it. If necessary, tie it off with a soft string (a shoelace is fine in an emergency) about two or three inches from the kid's body, and snip it off on the doe's side of the knot with a sharp scissors.

One of the most important accessories of the midwife is a good supply of clean cloths: lots of them! Wipe the kids off, paying special attention to make sure the mouth and nostrils are clear of mucus. The doe will gladly help by licking her newborns.

The umbilical cord must be disinfected. Iodine spray is convenient, but better protection can be had by pouring some iodine into a small container, pressing it up to the kid's navel, then briefly tipping the kid over to insure good coverage of the entire navel with the iodine.

Watch for the afterbirth. If the doe doesn't expel it within several hours get a vet: a retained afterbirth is nothing to mess around with. If it's just hanging out of the doe, don't pull on it. You might cause hemorrhaging.

When the afterbirth is expelled, dispose of it. Some does eat it, a natural instinct most wild animals have developed to protect their young from predators that might otherwise be attracted. Most goatkeepers don't care for this little ritual, but it won't hurt the goat. Won't do her any special good though, either.

ATTENDING TO THE DOE

Put the kids in a clean, dry, draft-free place (a large box works well) and turn your attention to the doe. She has lost a tremendous amount of heat, even in warm weather. Offer her a bucket of hot water. Most pet-type goats also get a special treat at this point: perhaps a small portion of warm bran mash or oatmeal, or a handful of raisins. Provide some of the best hay you have: she's earned it.

UNWANTED KIDS

Examine the kids. Remember that only the very best bucks from outstanding dams and sires should be kept for breeding—one mature buck can breed a hundred does a year, and many bucks are kept for five or more years. Mathematically, that means that less than one in a thousand bucks is required as a sire. If you don't intend to keep the bucks for breeding, you may want to use them for meat.

Check for supernumary teats (extra teats). There are several variations of this condition, some of which make the animal worthless as a milker. If the extra teat is sufficiently separated from the main two, it might not interfere with milking, and can even be removed at birth with a surgical scissors or by tying a fine thread very tightly around the base and letting it atrophy. Actual double teats make the animal worthless. Bucks can also have double teats, and such animals should not be used for breeding.

If the tip of the vulva on doe kids has a pea-like growth, the animal is a hermaphrodite and will not breed. It should be destroyed, and the mating that produced it should not be repeated. Some hermaphrodites are not obvious.

If you do end up with unwanted kids—bucks and does—they can be butchered and dressed like rabbit.

KEEPING THE KID WARM

Perhaps one of the most common kidding problems most people encounter is entering the barn on a blustery morning in late winter or early spring to find a newborn kid cold and shivering. If it seems to be doing all right, don't feel sorry for it and bring it in out of the cold. Place it in an enclosed box or pen padded with an old blanket or feed sacks, away from any hint of a draft, and with a heat lamp if the weather is really nasty. But don't let it get hot: the switch back to normal temperatures will be as dangerous as the cold that brought on the problem in the first place.

In the case of a severely chilled kid on the brink of death, more drastic action will be required. If you find one still wet and thoroughly chilled and nearly lifeless, one way to save it is to submerge it up to the nose in a bucket of water at about 100 degrees F. That's about the environment it just came from. When it has revived, dry it thoroughly, wrap it in a feed sack or blanket, put it in a box in a protected place, and watch it carefully.

However, such a kid might also be suffering from hypoglycemia, or low blood glucose. As the glucose level falls, the kid shivers, arches its back, its hair stands on end, and it moves stiffly. Eventually it lies down, curls up, becomes comatose, and dies.

The remedy is to warm the kid and administer at least 25 ml of 5.0 percent glucose solution with a small rubber stomach tube. When the kid is showing signs of reviving, get two ounces of colostrum into it, with the stomach tube if necessary. Return it to the barn as soon as it's active.

If you do end up taking a kid into the house for any reason during cold weather (and almost everyone does!), you're stuck with a goat in your house until the weather warms up. Even then you should harden it off by degrees rather than exposing it to the cold all at once.

COLOSTRUM

The kids will be standing and trying to walk on wobbly legs, perhaps within minutes of birth, and they'll soon be looking for their first meal. This must be their dam's colostrum, or "first milk," a thick, sticky, yellow, nonfoaming milk. Colostrum contains important antibodies and vitamins, and the survival of any newborn mammal is in jeopardy without it. It's so important, in fact, that if you buy a goat just a short time before kidding and she has no opportunity to build up antibodies to your particular location and therefore can't pass that protection on to her kids through colostrum, you might possibly encounter problems with sickly kids.

You can allow kids to nurse, or you can milk out the doe and feed the kids from a pan or bottle. It's extremely important to get some colostrum into them within half an hour or so of birth.

If you milk the doe, refrigerate the colostrum and feed it later. Warm it carefully, preferably over hot water: colostrum scorches much more easily than milk does.

Kids should get at least three feedings of colostrum a day for the first two days. By the third day, their ability to absorb colostral antibody is all but gone, and by then the doe is beginning to produce milk, not colostrum.

If for some reason the doe has no colostrum, you can make a fair substitute. Use 3 cups of milk, 1 beaten egg, 1 teaspoon of cod liver oil, and 1 tablespoon of sugar.

While generally considered not fit for human consumption, there is a pudding (called *beestings*) made from colostrum that some people consider a great delicacy. Most goat owners won't have enough colostrum to worry about making use of it. If you have extra colostrum, freeze it for future emergencies. Pour it into an ice cube tray, and when it's solid, turn the cubes out into a plastic bag. One cube is just right for one feeding.

CHAPTER 12

Raising Kids

After the excitement of freshening is over, your goat barn can settle into a routine. For the first three or four days your doe will produce colostrum, the thick yellow milk so necessary for the kid's well-being. After that there hopefully will be enough milk for the kids and you. And after two months of relative inactivity, your goat barn will be a hectic place. In addition to the usual feeding and cleaning tasks, you'll be milking twice a day...and raising kids.

Raising kids requires some knowledge and a lot of work and time. Different goat raisers have different opinions about how the job should be done, but none can deny that the first year of the goat's life—along with her breeding and prenatal care—is an important determinant in how she will behave and produce later.

EARLY FEEDING

If the doe has a congested udder or a very hard udder, the condition often can be helped by letting the kids nurse for the first few days. The suggested procedure is to bring the kids to their dam every few hours rather than leaving them together. This entails more work but it eliminates a lot of commotion and consternation later on. First fresheners, which often have very small teats, are also frequently left with their kids if the milker's hands are too large. The teats will enlarge with time.

If you milk the doe, do it within half an hour of kidding and offer the kids some colostrum. Feed milk at close to goat body temperature, which averages around 103 degrees F. Colostrum scorches easily: use a water bath or double boiler to warm it.

How will the kids be fed? Nursing is certainly the easiest method, but not necessarily the best, so far as goat breeders are concerned. Some people say it ruins the dam's udder, which is important not only if you intend to show her but also if you want her to have a long and productive life as a milker. Possibly a more important consideration is that you don't know how much the doe is producing or how much the kids are getting. Also, kids left with their mothers are much wilder than hand-raised kids. Another important consideration is that once a kid learns to suck its dam it will be difficult—maybe impossible—to teach it not to. Some does wean their kids relatively early, but there have been other cases where yearlings are still sucking and the milk supply for the kitchen is lost. The only solution in those cases is complete separation. It's better to do it right away.

PAN FEEDING AND BOTTLE FEEDING

Kids not left with their dams can be pan fed or bottle fed.

Many breeders prefer bottle feeding because it's more "natural." They point out that with pan feeding the animal is forced to lower its head to drink and milk can get into the rumen, where it doesn't belong. Digestive upsets can result. Of more immediate practical concern, pans or dishes can get stepped in...tipped over...and Nubian kids end up with milk-sopped ears, which can result in skin irritation.

On the other hand, pans are certainly much easier to fill, wash, and sterilize than bottles and nipples. You won't need lamb nipples or bottle brushes.

A variation on bottle feeding that's very popular where large numbers of kids are fed is a large container (such as a five-gallon pail) with special nipples. The nipples are attached to plastic tubes that reach to the bottom of the container. The kids suck on the nipples and the milk is drawn up through the tubes, just like drinking through a straw. Commercial units go by such names as Lam-Bar and Lamb-Saver.

If you're feeding more than a few kids with bottles, bottle racks are handy. You can feed as many kids as you want to at one time, without holding the bottles in your hands. You can buy plastic bottles with lamb nipples and wire racks that attach to board fencing, or you can easily construct a rack to hold soda or beer bottles. In a 1 x 4 of any length you want, cut or drill holes large enough to admit the necks of the bottles you'll be using. Nail this to a 1 x 10 or 12 so that the bottles will rest on the larger board, their necks and the nipples poking through the holes. Fasten the rack to a fence at a convenient height depending on the size of the kids, at about

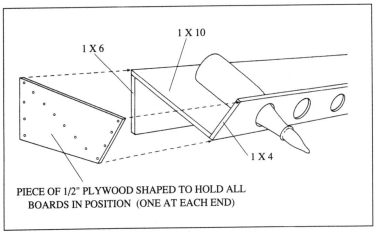

1 X 6

1 X 10

1 X 4

PIECE OF 1/2" PLYWOOD SHAPED TO HOLD ALL
BOARDS IN POSITION (ONE AT EACH END)

This handy bottle rack can be hung from the barn wall.

a 45-degree angle.

Hungry kids butt udders (and bottles), and when they get large enough, you'll probably need another device to hold the bottles more firmly in place. And occasionally a kid will pull a nipple off a bottle, dumping the milk. But overall we've found this rack to be a great time- and labor-saver.

Whichever feeding method you decide to use—pan, bottle, or nursing—once you start it you're stuck with it. It's difficult to teach a kid to drink from a bottle once it's used to a pan, and the other way around.

FREQUENCY

As for any baby, frequent small feedings are better for goat kids than infrequent large feedings. For the first three days the kids should get four to eight ounces of colostrum four times a day, depending on size and appetite. Then, until they're eight weeks old, they get anywhere from eight to ten ounces three or four times a day and an equal amount of warm water afterwards, if they want it. Most will. As the kids learn to eat hay and grain, the milk should be gradually decreased, and in most cases they should be weaned completely by the time they're eight weeks old.

Some people feed milk much longer—as long as six months, in some cases. This certainly isn't necessary, and according to Dr. Leonard Krook of Cornell University, it may actually be harmful. Kids overfed calcium (milk is high in calcium) are likely to develop bone troubles in later life.

MILK REPLACER

If you want all the milk your goat produces for yourself, the kids can be fed milk replacer. However, be sure to use milk replacer made for sheep or goats, not cows. Calf milk replacer is not high enough in fat, and goats will not do well on it. In fact, a large commercial dairy that has tried several brands of milk replacers and keeps meticulous records on all its goats claims that even when the kids looked fine while on replacer, two years later most of them were dead or had been culled because they lacked stamina.

Yet many goats are raised on milk replacers. There are very few hard and fast rules about anything connected with kid raising. This is one area where the old saying, "The eye of the master fatteneth the cattle" is especially true. You obviously don't want to starve the kids. But don't kill them with "kindness" either. This happens most frequently by overfeeding milk and causing scours (diarrhea), which can be fatal. You don't want kids to be "fat and healthy" because fat *isn't* healthy for a dairy animal. Strive for condition, not overcondition. The kid should be producing bone, not fat, to develop her full potential in later life.

WEANING

Early development of the rumen is extremely important for later production. Most kids will start to nibble at fine hay by the time they're a week old. They should be encouraged to do so with kid-size mangers and frequent feedings of fresh, leafy hay.

At weaning, most breeders feed a calf starter ration: half a pound, twice a day. At six months the kids are switched to a milking ration. By seven months doelings weigh seventy-five to eighty pounds and are bred.

Milk-fed kids weighing twenty to thirty pounds are in great demand as meat in some localities, especially at Easter and Passover.

Any bucks kept for meat should be castrated by the time they're eight weeks old. Bucks kept for meat require no special diet, but some chevon (goat meat) aficionados claim milk and browse produce the best meat. Butcher kids can be fed grain liberally.

————————————— CHAPTER 13—————————————
Milking

The new goat raiser must learn about goat feeds and nutrition, about bucks and breeding and raising kids, all for one purpose: to get milk. Milking, therefore, is at the apex of the pyramid of all goatkeeping skills.

Goats must be milked at regular twelve-hour intervals and according to a regular routine, for best results. Milking at 6:00 A.M. one day and 9:00 A.M. the next is one of the easiest ways to depress milk production. You might milk at 7:00 A.M. and 7:00 P.M. or at noon and midnight, but it should always be as close to twelve hours apart as possible, and always at the same time. (See Appendix A for an explanation of this.)

SANITATION

Of even more importance than regularity is sanitation. One of the main reasons for keeping goats is having milk better than any to be found in the supermarket dairy case. This requires not only a knowledge of dairy sanitation but a rigid adherence to sanitation principles.

EQUIPMENT

Milking equipment can be simple or elaborate. You could, if you wanted to, milk into a bowl from the kitchen cupboard, make a milk strainer from two inexpensive funnels, and store the milk in the refrigerator in fruit jars. At the opposite extreme you could spend hundreds or even thousands of dollars on milking equipment.

If you intend to milk 730 times a year for the foreseeable future, you will get more satisfaction and better quality from proper equipment.

A four-quart seamless stainless steel half-moon hooded **milking pail** is no luxury, despite its fairly high price (about $40, in 1989). Many of these have been in daily use for twenty years and more, which brings the per-use cost down to a pittance. Being seamless and stainless, these pails are easy to clean and disinfect. The hood and handle are removeable to enable thorough cleaning. They're made especially for goats, naturally, so you'll have to check the goat supply houses currently advertising in goat publications to find one. This is the one piece of goat equipment I couldn't do without.

Whatever you use, avoid plastic. No amount of cleaning can get the bacteria out of the pores in plastic, and you'll soon end up with a product that's unfit for human consumption.

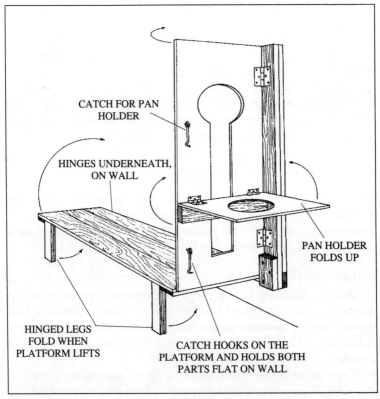

CATCH FOR PAN HOLDER

HINGES UNDERNEATH, ON WALL

PAN HOLDER FOLDS UP

HINGED LEGS FOLD WHEN PLATFORM LIFTS

CATCH HOOKS ON THE PLATFORM AND HOLDS BOTH PARTS FLAT ON WALL

A convenient milking stand that folds out from the wall.

A **strainer** is a necessity. It must be of a type that uses disposable milk strainer pads, available at any farm supply store. (I won't even comment on the practice of running the milk through a hunk of cheesecloth.) The strainer must be small enough to fit into your holding container, such as a wide-mouth canning jar. Small tin kitchen strainers are inexpensive and work well with small amounts of milk, but larger sizes made especially for backyard dairies are available in both stainless steel and aluminum. (Stainless steel costs about twice as much.)

If you're on a very tight budget, you can make a strainer from two large kitchen funnels. Cut off the spouts, and a little more, so you have two funnels with openings of 2-3 inches. Put a milk filter pad into one funnel, and place the other funnel on top to hold it down firmly.

Milk can be stored in one-, two-, or four-quart glass jars, or in aluminum or stainless steel cans of the same size. Again, avoid plastic if it's to be reused. Look for something easy to clean and sterilize.

In addition to these tools, the home goat dairy will require a bucket to hold udder wash; an udder sponge or cloths; cloth or paper towels for drying the udder; and udder wash and cleansers and disinfectants for utensils. A **scale** for weighing milk so you can record production isn't an absolute necessity, but it's a very good idea.

A **strip cup** will help you maintain a constant check on one aspect of your herd's health and the quality of the milk your family drinks. A strip cup is used to detect mastitis. This is simply a cup, with either a screen or a black tray at the top. You squirt the first stream of milk from each teat into the cup, and examine it for flakes, lumps, and other signs of abnormal milk.

FACILITIES

Ideally the milking should be done in a special room, away from the goat pen and any sources of dust and odor. It should have good ventilation, running water, electricity, a drain, a minimum of shelves or other flat surfaces that gather dust, and impervious floor, walls, and ceiling so it can be kept even cleaner than your kitchen. All this—and more—is required for commercial Grade A milk producers.

Ideals are often unattainable. Many people with just a few goats milk right in the barn or shed (not in the pen).

Goat milk actually has less bacteria than cow milk as it leaves the udder. But on the debit side, the goat milk is more likely to pick up coliform bacteria during the milking process. This is due in part to the dry nature of goat dung. It actually becomes dusty. Combined with the loose housing most commonly used for goats, this results in dung dust and coliform bacteria in the air, on flat surfaces, and on the goat. And the goat milker is more likely to disturb the hair on the belly than the cow milker just because of the size of the animal.

Whether milked in a separate parlor or in the goat barn, the goat should have long hair around the udder clipped, she should be brushed to remove loose hair and dust, the udder should be washed with a dairy disinfectant, and a hooded milking pail should be used.

A milking stand is far more comfortable than squatting, especially if you have a number of milking does, or if you tend to creak anyway. Milking stands have stanchions to lock the doe's head in place to help control her while you're milking, and a rack to hold a feed pan to keep her occupied.

MILKING PROCEDURE

Finally, we're ready for the actual milking.

It looks easy...until you try it. But with a little practice it *is* easy, and you'll wonder why you had milk up your sleeves and all over the wall and your legs the first time you tried.

Position yourself at the goat's side, facing the rear. Goats can be milked from either side, but they develop a definite preference for the side they're milked from. If you're right-handed, the right side will probably be easier.

WASHING UDDER

The first step is washing the udder with warm water and an udder-washing solution, available at farm supply stores. Follow the directions for the proper strength. Strong solutions can cause udder irritation.

Dry the udder and your hands with a paper towel—a fresh one for each goat—to avoid chapping and other udder problems.

DRAWING MILK

Then grasp the teat with your thumb and index finger encircling it near the base of the udder. Do not grasp the udder itself, which is sometimes tempting on goats without a clearly defined teat. That could cause udder tissue damage and the tissue can work its way into the teat with disastrous results.

Squeeze your thumb and index finger together to trap milk in the teat. This must be held firmly, or when you squeeze the rest of the teat the milk will be forced back up into the udder rather than into your pail.

Next, gently but firmly bring pressure on the teat with your second finger, forcing the milk down even further. The third finger does the same, then the little finger... and if all has gone well the milk has no place to go but out of the teat. Not necessarily where you want it, on your first try, but at least out of the teat.

The first squirt from each teat should be directed into a strip cup, a cup with a sieve or a black plate for a cover. That first stream is high in bacteria that have collected in the teat orifice and it shouldn't go into your pail. In addition, use of a strip cup will enable you to see any abnormality in the milk, such as lumps, clots, or stringiness. This is an indication of mastitis, which demands your attention (see Chapter 8). Never use the milk from any animal that's not in perfect health.

Then start milking into your pail, using first one hand on one teat, then the other hand on the other teat. With a little practice you'll develop rhythm.

Keep it up until you can't get any more milk, then massage or "bump" the udder, as kids do when sucking. You'll be able to get more milk. This massaging is important, not only because the last milk is highest in butterfat, but also because if you don't get as much milk as possible, the goat will stop producing as much as she is capable of.

The final step is stripping, or forcing out the very last of the milk in the teat. (You never get all the milk in the udder.) This is done with the crotch of the thumb, or the tips of the thumb and forefinger. Grasp the teat at the top again, and force the milk out by running your fingers down the length of the teat.

PROBLEM MILKERS

There are problem milkers. If you've learned to milk with decent animals, you can probably figure out how to cope with the other kind, but if you're unfortunate enough to have to learn on a troublemaker, some of the fun will go out of the experience.

First fresheners are most liable to be the culprits, although older does sometimes develop ornery habits, especially when they know you're using them to practice on. First fresheners are also likely to have small teats, which makes milking difficult, especially if you have large hands. On some it's possible to milk by using the crotch of the thumb; others will require using the tips of the thumb and index finger in what amounts to stripping.

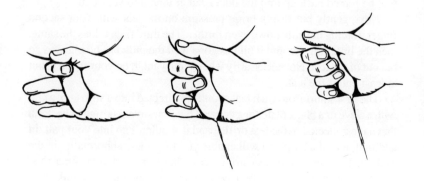

Left. *Milk can run out of the teat into the pail or back into the udder, so first close your thumb and first finger so the milk cannot run back into the udder.* Center. *Next close your second finger—and the milk should squirt out. Discard the first stream—it will be high in bacteria.* Right. *Close the third finger. Use a steady pressure. Don't jerk down.*

An occasional doe will have a tendency to kick, and almost any doe might kick once in a great while. This generally indicates that something is wrong. She's bothered by lice or flies, or you pinched her, or your fingernails are too long. Placing the bucket as far forward as possible (away from her hind legs) will help in this situation, and you can lean into her with your forearm to control movement. Leaning into the goat with your shoulder, holding her against the side of the milking bench or wall, will also serve to restrain ornery or nervous animals. It can also be useful in mild cases of "lying down on the job." Goats that have been nursing kids are especially prone to this sort of unhelpful behavior.

Goats can be milked by machine, but since goat-milking machines can cost over $1,000, we won't discuss them in a book for beginners.

CLEANING EQUIPMENT

All equipment that comes into contact with milk must be scrupulously clean. Milk is not only a highly perishable product (especially raw milk), but it is also extremely delicate. Milk must be cooled *immediately* after milking. It should not be left standing while you finish chores. Small containers may be cooled in the refrigerator, but anything more than a quart

Upper left. *Next close your little finger . . . squeeze with whole hand.* Upper right. *Now release the teat and let it fill up with milk. Repeat the process with the other hand .* Center left. *When the milk flow is near to stopping, nudge the bag to see if the doe has let down all her milk.* Center right. *The final bit of milk may be stripped out. Take teat between thumb and first finger.* Lower left. *Now run down length of teat. Milk high in butterfat usually comes at end of milking, but prolonged stripping is bad for the teat and udder.* Lower right. *The first milk is milked into the strip cup. If the milk is "lumpy" it will not pass through the strainer.*

will not cool rapidly enough for good results unless it's immersed in ice water. Ideally, milk should be cooled down to 38 degrees F. within an hour after leaving the goat. That's quite a rapid drop when you consider that it was over 100 degrees F. when it left the udder. Home refrigerators aren't cold enough.

Cleaning milking utensils is quite different from ordinary household dishwashing. A dishcloth will not clean microscopic pores that hold bacteria that will spoil milk or give it a bad flavor: a brush must be used. Household soaps and detergents contain perfumes that will leave a film on equipment and may cause off-flavors in the milk. Household bleach isn't pure enough for the dairy. You can't even use tap water to rinse milking equipment, or towels to dry it, because of the bacteria these contain. You need special compounds.

There are four dairy cleaning agents. Two are for washing: alkaline detergents and acid detergents. Iodine and chlorine compounds are used for sanitizing.

The alkaline detergent is the basic cleaning agent. However, it leaves a cloudy film called milkstone, which harbors bacteria. To get rid of the milkstone you must use an acid detergent, which doesn't have the cleaning power of the alkaline detergent. Most dairy farmers scrub their equipment with alkaline detergent for six days, and on the seventh, when everybody else is resting, they scrub with acid detergent. If you have hard water that makes milkstone develop faster, you can put the acid detergent in the rinse water every day.

Chlorine compounds are used to sanitize equipment, but they're too strong to use for washing udders. There you need iodine compounds. Actually, iodine compounds can also be used to sanitize equipment if you measure carefully and let the equipment soak long enough. At least five minutes is required. Follow the directions on the label of the product you use.

Measure any of these materials carefully. If the solutions are too weak, they won't do the job they were intended for; if too strong, you're wasting money and you run the risk of contaminating your milk.

Never let milk dry in a pail or other piece of equipment. Rinse it as soon as possible with cold water, then wash in warm—not hot—water and dairy detergent. Rinse in plenty of warm water with sanitizing compound and invert on a rack—not a shelf—to air dry. Do not use a towel.

Commercial dairies *must* follow these procedures. The number of backyard goat raisers who go through all this is open to question. I've described them not because you'll croak from drinking milk that wasn't produced under hospital conditions, but so those who want to do a professional job will know they aren't doing it with soapy dishwater and a dishcloth and towel. If you ever encounter "bad" tasting milk, your milk handling and equipment cleaning procedures are the first things to examine.

CHAPTER 14

Keeping Records

Recordkeeping is necessary for the commercial goat dairy, because only through accurate and complete records does the owner know if the operation is making a profit...and if not, why not.

Recordkeeping is a necessity for the show goat breeder because only accurate and complete records will help to upgrade goats to the hoped-for blue ribbon status.

Most homesteaders and other backyard goat raisers shun records because they aren't involved with profit or upgrading or awards, and they think the work is a boring waste of time.

They're wrong, on three counts.

It's true that the home dairy doesn't depend on goats for a living, as the commercial dairy does. But profits (and losses) show up in milk and dairy products that are better and cheaper than those purchased in the supermarkets. Even if the casual goat owner has no intention of ever entering a show ring or even coming close to a goat show, it's still necessary to know certain facts about the herd's production, and the results of management and breeding practices.

And recordkeeping can be fun! It becomes a challenge to have does that produce better than their mothers, and it's satisfying to look back on records that are several years old and see, in black and white, how you've progressed. No livestock breeder of any kind can afford to be without good records to use as a management tool.

If you have registered goats, pedigrees and registration certificates will be an important part of your files. The person you buy registered goats from will help you get started with these. There are several registries, with slightly different procedures: get information on specifics from whichever one you choose to work with.

BARN RECORDS

Of more interest to most beginners, perhaps, are barn records. The basic barn record is a chart showing how much milk each goat produces. A plain sheet of paper with the goats' names written across the top and the days of the month down the left margin works fine. You can write the morning's milk in one corner of each square or imaginary square, and the evening's milk below it, as 4.5/4.

Milk is measured by weight rather than volume for official records on goats and cows. It's the best procedure for the home dairy, too. Freshly drawn, unstrained milk foams, and it's difficult to gauge actual production in quarts, pints, or even cups. Then too, it's much simpler to deal in pounds and tenths of pounds rather than in quarts and fractions of quarts. For all practical purposes, a quart of milk weighs two pounds: eight pounds is a gallon.

It's a good idea to use this sheet to make notations of relevant data. For example, if you note "Susie in heat," you will be alerted to watch for her next cycle in twenty-one days. Notes on changes in feed, unusual conditions such as a doe not feeling well or acting quite right, or any other factor that might contribute to differences in milk production can be a big help in interpreting your records even years later.

It's convenient to note breeding dates and the name of the buck, and freshening dates with all pertinent information, right on this same sheet. When we bought feed instead of growing our own, that went on the sheet, too. At the end of the year we had a complete record of the input, output, and interesting happenings in our barn, all on twelve pages.

RECORDING REALITY

One of the primary advantages of such a system is that it overcomes the natural forgetfulness of most human brains. Let's face it: few people, if any, are going to remember the statistics from 730 trips to the barn a year which, if the herd consists of three goats with lactations of 305 days, means 1,830 separate entries each year for milk alone.

In addition to not being able to remember all those numbers, the brain can distort them. For example, you might be impressed by Susie's production of a gallon of milk in one day and consider her the best goat in your herd. But your records might show you that a less spectacular producer that just chugged along less dramatically, but with a long and steady lactation, actually produced more than the flashing star. If you had to cull milkers or make a decision about whose daughter to keep, you might make the wrong choice without records.

In many cases breeders will note that a relatively few top does produce as much as a larger number of poorer producers. Since poor producers require just as much work as good ones and eat virtually as much, it follows that milk from the lower third or even half of your herd costs more than milk from the top half or two-thirds. This suggests that you could get more milk for the same amount of time, effort, and money by replacing the poor doers with daughters from the best does. If you don't need that much milk, you might be able to eliminate one or more animals, reduce your feed bill, and still increase your milk production.

Records of breeding, expenses, income, and milk production are all basic, and it doesn't take much time or knowledge of accounting to keep them. Just the act of writing them down will tell you a few things about your operation, but it's also possible to squeeze a lot more helpful and interesting information out of those records. (A few enterprising individuals are already offering computer programs for goat breeders. Since this is a very new and rapidly changing field, any information given here would soon be out of date: check for ads in the current goat periodicals.)

BALANCE SHEET

Of course, you don't need a computer to tell you how much your goats are worth to you. You can keep monthly and annual tallies of a few simple income and expense items such as those listed below.

We didn't provide examples of specific prices for several reasons. In any given category, your cost and income will vary according to your location, your type of operation, and your management methods. Even when those factors remain constant, your cost can vary from year to year. During the drought of 1987, hay prices doubled (and more) in many places. You might have mostly doe kids one year, with accompanying heavy registration expenses. The next year you might have mostly bucks that go to the butcher and very low registration expenses. Medical expenses often come in spurts.

In addition, nationwide averages would be meaningless here, even if they were available. There's no point in looking at numbers and setting unrealistic goals for yourself—or being disappointed because you think you don't measure up—or saying this book is way off base because its numbers don't jibe with yours!

Actually, it doesn't matter what anyone else's income and expenses are: yours are going to be different, and it's *your* bottom line you're concerned about. A balance sheet like this will tell you if your herd is profitable or not, and how much it made or lost.

Income
    Milk sales   _____
    Sales of stock   _____
    Family milk   _____
    Stud fees   _____
    Boarding fees   _____
    Miscellaneous (Disbudding services, etc.)   _____

    Total   _____

Expenses
    Purchase of stock   _____
    Feed—grain, minerals, salt   _____
    Hay   _____
    Veterinarian and medicine   _____
    Repairs on equipment   _____
    Supplies   _____
    Advertising   _____
    Registry, transfers, etc.   _____
    Telephone, postage   _____
    Amortized costs   _____

    Total   _____

    Profit/Loss (Income minus expense)   _____

This balance sheet tells us if the herd is profitable and how much it made or lost. If it lost money, we want to know why, and what can be done about it. For that we must go to other records. As an example, let's look at the notes made on a small herd of Nubians.

EVALUATING LOSSES

In the first place, total milk production for the six does in this herd was 5,900 pounds. That's a very poor average. Two does came in with little milk and were dried off after a few weeks, but were kept on the payroll.

There may have been good reasons for their lack of production, and maybe there was a reason for keeping them, but statistically speaking they should have been culled. The bottom line of the balance sheet would have come out looking much better. In fact, just this one simple step could have turned the loss into a profit.

The records show that this was a poor kidding year. Eight bucks and four does were born, and one of the does died. Over the years any given herd will produce bucks and does pretty much on a 50/50 basis, and the sale of additional doe kids could have helped the balance sheet. Kid sales, particularly where the kids are valuable and there's a good market, frequently mean the difference between profit and loss.

A good chunk of money went for medical expenses. Checking into the nature of the problems might indicate a condition that could have been corrected by different feeding or management practices, or if the vet bills apply to one or two does or to a particular family, perhaps culling is in order.

Fencing and the purchase of stock should rightfully be treated as capital expenses. These aren't used up in a single year, and their cost should be apportioned over their useful lifetimes. Other examples of expenses that should be amortized would be milk pails, strainers and stands, barns and pens, and metal cans for grain storage.

If you're an accountant, you might enjoy or feel obliged to go into great detail; if you hate numbers and seldom even get your checkbook balance to agree with the bank statement, you'll want to make all this as simple and rudimentary as possible. The important thing is to *do* it, and that means using a system you can be comfortable with. Having these records, and using them, is the only way to make intelligent management decisions. It's the only way to lower your milk bill.

Your records can tell you what each of your goats is worth. Just add up her annual milk production, put a price tag on it, and add any income from kids. "Income" might mean cash received, or it might be in the form of meat, or a replacement doe.

PRICING THE MILK

Pricing the milk from your home dairy is no simple, cut-and-dried calculation. Even if you go by the maxim that anything is worth only what someone else is willing to pay for it, goat milk presents special problems.

If you have a baby who's allergic to cow milk and can't find goat milk—at any price—the cost of your home-produced milk is probably of little concern.

If you're feeding milk to pigs or calves—or dumping it on the ground— the value of the milk is obviously much less.

Most of us produce milk that has varying value. In winter, when production is likely to be low, the entire output might be used for drinking (the "fluid milk" market, the dairy cow industry calls it). If we'd be willing to pay the health food store price of goat milk, this milk is quite valuable. If, without goats, we'd be drinking cow milk, it's somewhat less valuable.

As the milk flow increases, we might begin to use some to make cheese or yogurt. This is "manufacturing" milk, and even cow farmers get less for it.

When we begin to use an even greater surplus as feed for pigs, calves, or puppies, the value slides even further.

In other words, you have a lot of leeway in putting a price on the milk you use yourself. Remember, you're not keeping these records or coming up with numbers for the bank or the IRS: they're strictly your own, to be used to improve your herd's performance.

CAPITAL COSTS AND OPERATING EXPENSES

Figure your capital costs, that is, money you spent on things that aren't used up all at once. This includes milking equipment, feed pans and water buckets, fencing, tools such as the disbudding iron and clippers and tattoo set, and the goat itself. Naturally you don't want to charge all this against the milk produced in one year.

The milk pail might last twenty years: take one-twentieth of the price as this year's cost. The goat might be good for another five years: take one-fifth of what you paid for her. Go down the list of capital goods, determine the capital costs for one year, and you'll have a more honest picture of your true costs.

Then add up your operating expenses: hay and grain, electricity used in the barn, veterinary fees, milk filters, and everything else that was purchased and used up.

By adding up the operating expenses and one-year cost of capital equipment and stock, and subtracting that from the value of the goat's production, you'll have a pretty good idea of her annual value to you. By adding up all these costs and dividing by the number of quarts of milk produced, you'll know the actual cost of your milk.

Even this isn't completely accurate, but it's adequate for most people, and far better than a complete disregard for accounting. If you're inclined, you can figure in the cost (or value) of labor, the value of manure, the cost of money or the return on investment, and even more.

If more goat raisers kept such records, you can be sure there wouldn't be very many $10 or $20 goats for sale! More people would pay better attention to culling and proper management, too, if they knew what their goats were really costing them.

MARKET CONDITIONS

One last word on records: Many books (including the first edition of this one) have said that many herds break even not because of the value of the milk, but because of the value of the kids. It's apparent (it was said) that a purebred and registered herd that can command top price for its animals will come out far ahead of a herd of grades whose kids are a drug on the market.

This is still true, but notice the wording: "*can* command top price." Many readers have pointed out that at a particular time in a particular location, even excellent purebred stock doesn't bring high prices. Many factors are involved: for top prices you not only need top animals, but you have to earn a reputation in the show ring, you'll probably have to be on test and have your animals classified, and you have to advertise. All these require time and money. How can you know if it will pay off?

One way is to make projections based on current prices and costs and market conditions in your particular area. This is just one more way of making records work for you.

CHAPTER 15

Chevon

The meat of goats is called chevon. People don't eat ground cow, pig chops, or leg of sheep: beef, pork and mutton sound much more appetizing. So back in 1922 the Sheep and Goat Raisers' Association of Texas held a contest to find a trade name for goat meat that could compete with beef and pork (which come from the French *boeuf* and *porc*, cow and pig.)

The winner was chevon, from *chevre*, the French word for goat, and the *-on* from mutton.

Of course, we might also call goat meat *cabrito*, from the Spanish for "little goat," but except for in the Southwest, chevon is the more common term.

Chevon is very popular in certain cultures, in this country especially among people of Spanish, Greek, and Jewish heritage. Kid is an important part of the meals for spring festivals of Easter and Passover. While most meat goats are raised and slaughtered in the Southwest, where they can be raised cheaply on range and where a large market exists, more than 40,000 cull goats and kids a year are being sold in the Northeast to serve the large New York ethnic markets. (Cull does have been selling for around $50; kids have been between $20 and $40.)

In most of the country, dairy goats raised in confinement can't be considered meat animals because of the labor and expense involved in raising them: it simply isn't profitable. In addition, their growth rate is too low to make them efficient or economical meat animals. A ration that will make a good lamb gain 0.66 to 0.99 pounds a day will only result in a gain of 0.33 to 0.50 on a goat. This is largely a matter of genetics. Meat is, however, an important by-product of dairying. Over the years any farm will average 50 percent buck kids. Not one in a hundred can be kept, profitably. While there is a limited demand for wethers (castrated males) as pets in some areas, it is probably more merciful in most cases to butcher them for meat.

In addition to unwanted males, any dairy—cow or goat—will have cull or aged does that simply are not paying their way. Resist any temptation to sell them as milkers to someone else. You might make a few dollars on the deal, but the cost to the goat world—and to your reputation—will be far higher.

SLAUGHTER

Culling is a fact of life when dealing with livestock, but that doesn't make it any easier—especially for city people with no livestock experience, and even more so when dealing with animals like goats! However, once you overcome any initial reticence, you'll be likely to agree that chevon is a delicious bonus of your home dairy.

Goats are commonly slaughtered at one of four stages. Most popular for the Easter-Passover market are milk-fed kids weighing between twenty and thirty pounds. On farms where milk is valuable or where the labor required to raise kids is deemed out of proportion to the value of the meat, kids can be butchered at birth.

If you are or become fond of chevon, you can castrate buck kids and feed them out for six to eight months.

And finally, cull does can be processed into jerky, salami, or anything that makes use of meat that isn't especially tender. (Bucks that are being culled should be castrated about two months before butchering and fed a liberal grain ration.)

If you dispatch the animal with a gun, aim from the back so as not to frighten it by the sight of the barrel. Some people prefer to use a hammer: a sharp blow to the skull will render the animal unconscious, and the jugular can be cut with a sharp, stout knife.

Although some people claim it's merely a cultural practice, most will say that thorough bleeding is important. Do not damage the heart in the killing process so it can continue to pump blood, and hang the animal head down to allow complete drainage. If you do much butchering, a gambrel hook will prove useful, but a carcass can be hung by passing a metal or wooden rod through the tendons of the rear legs, or even by tying it to a rafter or a tree branch with a rope.

BUTCHERING

We've been told that the Greeks, who have a good deal of experience in goat butchering, cut a small incision between the hind legs and blow up the hide like a balloon. This helps separate the hide from the meat and makes skinning easier and cleaner. No doubt most people would rather use a tire pump than their mouths, but one goat raiser reported inserting the nozzle of a garden hose into the incision and filling the space with cold water. In addition to separating the hide, this helped cool the meat.

To skin the animal with or without this step, carefully cut a slit from between the hind legs to the throat. Don't cut too deeply: you don't want to cut into the intestines or meat. Once started you can usually work your fingers beneath the skin to hold it away from the body.

From the two ends of this cut, continue out along the insides of all four legs. The skin is tighter on the legs and again, try not to cut into the meat.

The pelt will be attached at the tail. To remove that, cut around the anus and loosen it until a length of colon can be pulled out. Tie off the colon with a piece of strong string to avoid possible contamination. Cut it off above the string and let it fall back into the body cavity. Then cut off the skin at the base of the tail.

If you're a very frugal person, skin out the tail and use it for stew meat.

If you're saving the hide, cut it off as close to the ears as possible. Skins from newborn goats are more like fur than hide, and many useful items can be made from any tanned goat skin. If you aren't interested in keeping the skin, you can cut the head off with it. In either case remove the head at the base of the skull.

Now cut down the belly, from between the hind legs to the brisket. If the animal has been starved for twenty-four hours before butchering (don't withhold water), the paunch will be empty and there will be less chance of cutting into it, but be careful anyway. Let the paunch and intestines roll out and hang.

Work the loosened colon end down past the kidneys and carefully remove the bladder. Pull out the liver and remove the gall bladder by cutting off a piece of the liver with it. If the gall bladder breaks and spills bile on the liver, wash the meat in cold water immediately to avoid a bitter taste.

The offal will fall free when the gullet is cut. Saw the brisket—an ordinary carpenter's saw will work if you don't have a meat saw—and remove the heart and lungs. Clean out any remaining pieces of tissue, wash

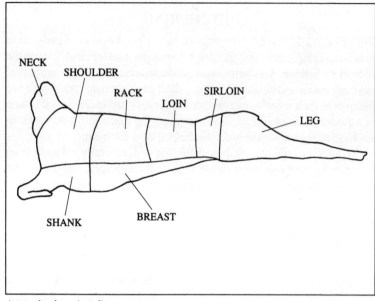

A standard cutting diagram.

the carcass with cold water, and wipe it dry.

The skull can be split to get the brains, and the tongue removed. Wash the liver, heart, and tongue in cold water and drain them.

Newborn goats weighing about seven pounds can be cut up like rabbits. Cut through the back just in front of the hind legs and again just behind the front legs. Each of these can be cut along the spine, giving you six pieces of meat.

Larger animals, cut like lamb, will yield roasts, chops, ribs, and trimmings that can be ground and used in patties or mixed with pork for sausage. Some of the larger pieces such as the legs can be cured like hams, or corned.

If you have little or no butchering or meat cutting experience, the thought of converting a carcass into neat packages like you see at the meat counter might seem like a formidable task. But don't let it throw you. Simply cut off pieces that "look about right to be a roast," and they'll be just as edible and probably even better-tasting than those from the supermarket. You really can't do anything "wrong" at this point.

If you enjoy lamb, use your favorite lamb recipes with chevon. Since we raise both sheep and goats we often have both in the freezer, and it's hard to tell the difference.

There are many ethnic dishes from such areas as Greece and Turkey that call for chevon. Oregano is a good spice to use with chevon, and the meat is excellent in curries. You'll find a few suggestions for getting started with chevon cookery in Chapter 17.

CHAPTER 16

Dairy Products

One of the joys of having goats is the dairy products you can make in your own kitchen. At certain times of the year you will have a surplus of milk that can be turned into a variety of products that will make you more independent of the supermarket and will make your goats more valuable to you. It doesn't make sense to produce cheap milk and then throw half of it out because you can't drink it all. And perhaps most important of all, making cheese and other dairy products can be very satisfying, and a lot of fun!

Milk can be frozen, and if you know you're going to run short later, maybe this is the first course to consider. Freeze it in plastic jugs, leaving an air space for expansion. Thawed frozen milk is somewhat watery, and while it's fine for cooking, you'll probably want to mix it with fresh milk for drinking.

Milk can also be canned. Use a pressure canner and approved pint jars and lids. Fill the jars to within one inch of the top. Put on the sterilized lids and rings, and process for 12 minutes at 12-1/2 pounds pressure (at sea level). Remove the jars from the canner, tighten the lids, cool, and store in a dark place.

The butterfat comes to the top and the calcium settles to the bottom in canned milk.

CHEESEMAKING

Sooner or later you'll want to try making cheese. In its most basic forms making cheese doesn't require much equipment, it takes little actual working time, and it can become quite a hobby...or even a business. There is as much art to making cheese, however, as there is to making wine. Don't expect to come up with any "vintage" cheeses without experience…or luck. (Neither of these helps me very much, but it's fun anyway, and the cheese is at least edible, usually.)

To get started with the most basic forms of cheese, you'll need rennet or junket, a dairy thermometer, and cheesecloth. These, as well as many other items, are available from several mail order firms (see Sources).

COTTAGE CHEESE

The easiest cheese to make is cottage cheese.

Start by warming one gallon of milk to 86 degrees F. Crush a quarter of a rennet tablet and mix with a small amount of cold water (or follow the directions that come with the rennet you use, liquid or tablet). Add this to the milk, stir, and let it stand in a warm place until a curd forms, usually an hour or so. When that happens (and the timing isn't critical) cut the curd with a long, thin-bladed knife. By cutting it in half-inch squares on the surface, and then slanting the knife 45 degrees and making similar slices at right angles to the first ones (reaching to the bottom and sides of your pot), you'll end up with small cubes.

Stir the mass, very gently, and cut any large pieces remaining. Slowly warm the curds and whey to 110 degrees F. This heating should be very slow and gentle. Within limits, the longer it heats, the firmer the curd will be.

When it reaches 110 degrees F. pour the curds and whey into a colander lined with cheesecloth and let it drain for a few minutes. Then run cold water over it to rinse off the whey.

It's ready to eat as is, or you can add salt, cream, chives, or anything that suits your taste. I hear this keeps a week in the refrigerator, but set on the table ours doesn't last nearly that long.

It's possible to make cottage cheese (and others) without rennet. Some recipes call for adding vinegar or lemon juice, cultures or a starter, buttermilk, sour milk, or even yogurt. (By the way, rennet is made from the stomach lining of a milk-fed calf or goat.)

In spite of the hundreds of cheese recipes available and the fact that I love to play in the kitchen, I usually stick with one basic recipe. One reason is that I like it. Another is that, although I've been using it for years, it seldom turns out the same twice! Cheese is affected by the length of time you heat the curd, how hot it gets, the amount and duration of pressing, and even such factors as humidity. It's difficult to make hard cheese during the summer in humid regions. Obviously I don't profess to be a knowledgeable cheesemaker: if I can turn out a tasty product, anyone can. Here's the basic recipe, the first part of which is the same as the cottage cheese we just mentioned.

BASIC CHEESE

Warm the milk to 86 degrees F. in a large kettle. It takes about 10 pounds of milk to make 1 pound of cheese, so I don't like to mess around with less than 6 quarts of milk, but that's up to you.

One-quarter tablet of rennet will suffice for any small amount of milk you might be using in a kitchen situation. Crush it with a spoon, add enough

Left: *Recipes call for cutting curd into curds of specific sizes. For half-inch curds, make half-inch slices.*

Right: *Then turn pot and slice again at right angles to first cut. Make certain to slice to bottom of pot.*

Left: *Finally, cut diagonally, all the way across the pot.*

water so it will dissolve, and add it to the milk with the heat turned off. Let it stand until a curd forms.

The curd is ready to cut when it's firm enough to be lifted up by a finger. Using a long knife (we have a bread knife, which seems to be ideal), cut the curd into half-inch squares. First hold the knife straight up and down and cut to the bottom of the kettle. Then cut at right angles to the first cut, at about a 45-degree angle. Do the same thing in the other direction, making sure to reach the bottom and sides. Stir it carefully to find any pieces you missed, and cut them so they're all about the same size.

Then reheat the cheese-to-be...*very* slowly. The temperature shouldn't rise more than 2 degrees every five minutes. Bring it up to 100 degrees, and hold it there until the curd is as firm as you want it. Stir it once in awhile to keep the curds from lumping together, but do it gently: you don't want to mash them. It takes 1-1/2 to 2-1/2 hours for the curd to get to the right consistency. If the curd is not firm enough when you remove it, the cheese will be pasty and sour. If it's too firm, the cheese will be dry and crumbly.

When it's ready, pour the curds and whey into a large container lined with cheesecloth. Lift the cheesecloth and let the curds drain by hanging it over the pan.

When it stops dripping, place the curd in a container and add salt to taste.

The whey makes good feed for chickens and pigs, but don't feed it to kids or goats. It will cause scours.

Notice that this is how we made cottage cheese. But this time we'll go further, and press more whey out of the curds.

You can spend a lot of money on a fancy cheese press, or you can construct one. It's easy to make a nice and very serviceable press by cutting the bottom from a three-pound coffee can or something similar. Cut two pieces of wood in circles that will fit inside the can. The top piece can be weighted to press the cheese down; the bottom piece will keep the cheese from oozing out the bottom of the can. The illustration here will give you some ideas, but many variations are possible.

You don't even need a regular cheese press for your first efforts. Merely shape the curds into a ball and make a bandage or headband from a clean dishtowel. Make it about two inches wide. Wrap it around the cheese and fasten it with a safety pin.

Place it on a board and lay another board on top of it. It will drain: put in a pan or kettle, or in the sink. Weight the top board with a brick or anything else that's heavy. Hard cheese requires quite a bit of pressure.

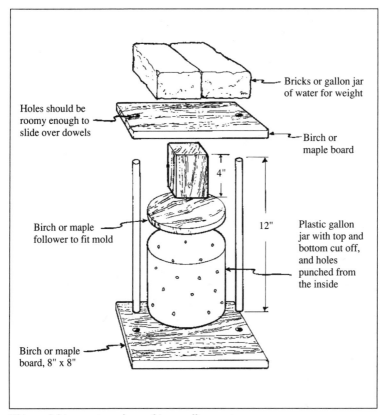

Bricks or gallon jar of water for weight

Holes should be roomy enough to slide over dowels

Birch or maple board

4"

12"

Birch or maple follower to fit mold

Plastic gallon jar with top and bottom cut off, and holes punched from the inside

Birch or maple board, 8" x 8"

This model is easy to make, and is equally easy to use.

Some cheese presses allow you to screw down the top to increase pressure; others have lever arrangements that allow pressure to be applied without stacking up 30 or 40 pounds of bricks...which tend to fall over when the cheese starts to settle, usually in the quiet middle of the night.

Turn the cheese several times a day. Generally speaking, the longer the pressing and the greater the weight, the harder the cheese. There are specific times and weights for various recipes, so you really can't go wrong in this regard, but you might want to record what you did in case you turn out a product so good you want to try to duplicate it. If you use anything handy for a weight and you turn it according to your own schedule and lifestyle, you might end up with a cheese that's different from someone else's, but it'll still be cheese.

When it has pressed for a few days, wipe it with a clean dry cloth and check it for cracks and pores. Seal these by dipping the cheese in warm water and smothing them over with your finger.

Put the cheese on a shelf in a cool, dry place. The ideal is said to be 56 degrees F.—too warm for the refrigerator and too cool for most other home locations, perhaps except for brief periods of the year in some areas. This is no insurmountable hurdle, and although it's one more factor that will affect the finished product, don't let it keep you from making cheese.

Turn it daily to keep the bottom from molding. In a few days a rind will begin to form. It should then be paraffined to prevent it from drying out completely. A half pound of wax is enough for a small cheese. Melt the wax, hold half the cheese in it, and let it cool. Do the other half and check your work for leaks.

The cheese goes back on the shelf, but must still be turned daily. It can be eaten as is, and you'll probably be anxious to taste-test your first attempt, but the length of curing also affects the finished product. A mild cheese will be ready in about six weeks; a cheese cured for six months will have a sharper flavor. Curing takes longer at lower temperatures, but if the temperature is too high, spoilage will occur.

As you can see, many variations are possible even with one basic recipe. By controlling your cheesemaking processes and keeping records of what you do with each one, you will probably find a method that suits you. Of course, if you really get into cheesemaking, the possibilities are endless.

For example, one form of cheddar can be made by following the basic recipe up to pouring off the whey. Then the cubes of curd are placed in a colander and heated to 100 degrees F. in an oven or a double boiler for 1-1/2 hours. When the curd forms a solid mass instead of the individual cubes you started with, slice it into one-inch-thick strips. Turn these every 15 minutes to allow them to dry evenly, still holding the temperature at 100 degrees. After an hour go back to the basic directions beginning with salting the curd. Cheddar takes six months to cure.

One of our favorite variations is a form of feta. When the basic cheese is pressed, instead of curing it, cut it into cubes about two inches square and place these into jars of brine. Eat them right out of the brine.

The possibilities are endless. True, there are certain cheeses that can't be duplicated exactly, at home: Limberger, Camembert, and certain others require special cultures that are closely guarded proprietary secrets. On the other hand, some cultures are available from cheesemaking supply houses. There are more than enough recipes and cultures to keep any cheese-loving goat owner busy, and happy.

YOGURT

Yogurt is another product that is popular among goat raisers. Home-made yogurt is so superior to the supermarket variety that there's no comparison. There are several ways of making it, with everything from commercially manufactured yogurt makers, to heating pads, to solar energy.

Warm the milk to 100–110 degrees F. Add culture. You can buy dried cultures, or use one cup of store-bought yogurt and save a cup of your homemade product to start the next batch. This will wear out after a while though, and you'll have to use a fresh culture again. Adding one-half cup of powdered milk to each 3-1/2 cups of goat milk will produce a firmer product.

A yogurt maker will automatically keep the milk at 100 degrees, but you can also put the warm milk and culture mixture into a preheated thermos and wrap it in towels to help hold the heat in. Or set a casserole dish in a warm oven and leave it overnight with the heat off. You can also use a heating pad, the back of the old wood cookstove, or on the right kind of day, set a glass-covered container in the sun. It will take five to six hours at 100 degrees.

Remember, yogurt is produced by "good" bugs: sterilize all equipment used in the process to eliminate "bad" bugs. Antibiotics in the milk will kill the yogurt-making bacteria.

BUTTER

Butter is difficult to make from goat milk, but only because of its "natural homogenization." If you have a cream separator there's no problem, but without one you'll have a hard time getting enough cream to warrant cranking up the churn.

(You might be fortunate enough to find an antique separator in operating condition at a reasonable price, but new separators cost hundreds of dollars and few owners of the poor man's cow can justify the expense.)

Lacking a separator, one method of getting cream from goat milk, or at least some of the cream, is to leave the milk in a flat pan with as much

An idyllic goat scene.

surface as possible exposed to the air. You'll be able to skim off some of the cream in about twenty-four hours. This cream will keep for a week in the refrigerator, so you can skim the cream from daily milkings for up to a week to accumulate enough to make butter.

The cream should be "ripened," which will happen naturally if it stays in the refrigerator for a week. Otherwise, leave it at room temperature for a day.

Put the cream into the churn, but don't fill it more than half full.

Don't have a churn? Then use an electric mixer, or a French whip, or just shake it in a covered canning jar.

The butter should "come" in about 20 minutes. Drain off the buttermilk. This isn't *cultured* buttermilk like you buy in the store (it's thinner), but it's drinkable and good for cooking.

Now you have to work the butter. Using a spatula or wooden spoon, press the butter against the side of the bowl and pour off the buttermilk that is pressed out.

Then wash the butter in cold water to get out any remaining traces of milk, which will cause the butter to spoil. Repeat the washing until the water comes off clear.

Most people prefer their butter salted. Add salt according to taste and work it in.

You might be disappointed to see that your goat butter is white. That's normal. If you want to make it yellow, use a special "Dandelion" butter coloring: nothing else will stick to the butterfat.

When you get really good at making cheese, yogurt, butter, and ice cream (see Chapter 17), and you use fresh delicious goat milk in custards, baking, and other cooking, you might find that you don't have enough milk left to drink! The solution? Simple! Go back to the beginning of this book... and get more goats!

Recipes for Goat Products

While avid goat enthusiasts know that goat milk tastes better than cow milk, so far as cooking is concerned there really isn't much difference: use goat milk in any recipe calling for milk.

When milk is plentiful, however, most people tend to use more of it, and that often means looking for recipes in which milk is a major ingredient.

Here are a few family favorites that readers have shared with *Countryside* magazine over the years.

Milk Soup

2 quarts goat milk	1 tablespoon powdered sugar
1 teaspoon salt	6 egg yolks
1/2 teaspoon powdered cinnamon	4 thin bread slices

Boil the milk with the salt, cinnamon, and sugar. Lay the bread in a deep dish, pour a little of the milk over it, and keep it hot but don't let it burn.

Beat the egg yolks and add them to the milk. Stir over low heat until thickened. Do not let it curdle.

Pour it over the bread and serve.

Devonshire Cream

Let milk stand 24 hours in the winter, 12 hours when the weather is warm. Then set the pan on a stove over very low heat until the milk is quite hot. Don't let it boil: the longer this takes, the better. When it's ready there will be thick undulations on the surface and small rings will appear.

Set the pan in a cool place, and the following day, skim off the cream.

In Devonshire, England, in the nineteenth century, this cream was used to make a very firm butter. But it was also considered a gourmet item in London, where it was eaten with fresh fruit.

Cream Cheese

First make a culture by adding 1/8 cake of compressed yeast to 1 cup of warm goat milk. Let it stand in a warm room for 24 hours, pour off half of it, and add 1 cup of warm goat milk. Pour off half every day, add another cup of warm milk, and in about a week the yeast flavor will have disappeared.

To make cream cheese, add the culture to two cups of warmed goat milk.

After 24 hours, heat the resulting curd over hot, not boiling, water for about 30 minutes to firm it. Press it gently in cheesecloth to remove the whey. Add salt and sugar to taste. Keep refrigerated until used.

Variations on Yogurt

(For a yogurt recipe, see page 157).

Bavarian Cream: For each 2 cups of yogurt add cool but unset Jello (your choice of flavors) made double strength.

Sherbet: Stir frozen juice concentrate (to taste) into yogurt and spoon into small plastic cups. Stick a plastic spoon into the center and freeze. These can be removed from the cups and, using the spoons as handles, eaten like popsicles.

Sour Cream: Spoon any amount of yogurt onto a clean cloth, draw up the corners, and hang it to drain for 3 hours or until it's as firm as you want it. This can be used in any recipe calling for sour cream, but it breaks down under heat. To use it in beef stroganoff or other cooked dishes, add starch or flour as a stabilizer, and heat and stir gently.

Sweet Cheese

Bring 1 gallon of milk to a boil. Add 1 pint of buttermilk and three well-beaten eggs. Stir gently.

When the curd separates, drain and press.

This makes a very delicious mild cheese.

Cultured Buttermilk

Add 1/2 cup purchased cultured buttermilk to 2-4 quarts of fresh goat milk. Stir and leave at room temperature for 12 hours. Chill and serve.

Save half a cup of this as your next starter.

Dry buttermilk cultures are available from cheese and dairy supply houses.

Goat Milk Fudge

Mix 2 cups of sugar, 2-1/2 squares baking chocolate, 1 cup goat milk, and 1/4 teaspoon salt in a heavy pan. Cook over medium heat, stirring constantly. Bring to 236 F. degrees on a candy thermometer, or to soft ball stage (a few drops dribbled into cold water can be formed into a soft ball).

Remove from heat, add 1 cup of chopped nuts, 1 teaspoon of vanilla, and about a tablespoon of goat butter. Beat until thick and creamy.

Pour into buttered dish and when cool, cut into squares.

Goat Milk Pudding

Mix 2 cups of goat milk, 1/2 cup sugar, and a pinch of salt in a heavy saucepan. While heating slowly, beat one egg. Add to mixture and bring to the scalding point, stirring constantly.

Dissolve 4 tablespoons cornstarch in 1/2 cup more milk and add this to the scalding milk, again stirring constantly. Stir until thickened, remove from heat, and add 1 tablespoon of goat butter and 1 teaspoon of vanilla.

For flavored puddings mix about 2 heaping tablespoons of sweetened cocoa powder with the sugar before adding the milk, or substitute brown sugar for white for butterscotch flavor.

This pudding can also be used to fill cream pies.

Vanilla Ice Cream, I

2 separated eggs
scant 1/2 cup powdered sugar
5/8 cup heavy cream

5/8 cup milk
a few drops of vanilla extract

Beat the egg yolks, sugar, and vanilla in a bowl. Bring the milk to a boil and pour it over the mixture, stirring constantly. Let it cool.

Whisk the egg whites until stiff and lightly whip the cream. Fold these into the mixture, which should be quite cold.

Whisk the mixture in a bowl. Pour it into shallow trays and put it in the freezer until it's slushy. Put it back in the bowl and whisk it again. (This keeps large ice crystals from forming.) Pour it back into the trays and put it in the freezer again.

After it's frozen, put it in the refrigerator for 30 minutes or so to soften it slightly.

This is easy, but it doesn't make very much.

Vanilla Ice Cream, II

4 cups cream
1 cup sugar

1/8 teaspoon salt
1-1/2 teaspoons vanilla

Warm (but do not boil) 1 cup of the cream over low heat. Stir in the sugar and salt until they're dissolved. Chill, preferably overnight.

Then add the other 3 cups of cream and the vanilla, put it in the cannister of an ice cream freezer, and proceed. This makes about 1-1/2 quarts.

Vanilla Ice Cream, III

4 eggs, lightly beaten
1-1/2 cups sugar
1/2 teaspoon salt
2 cups milk

2 cups light cream
1 tablespoon vanilla powder
    (see note below)
4 cups cold heavy cream

Combine the eggs, sugar, and salt in the top of a double boiler. Whisk in the milk and the light cream and cook over simmering water, stirring constantly. When the mixture thickens slightly, remove it from the heat.

Add the vanilla powder, straining it through a large-mesh sieve. Stir thoroughly and refrigerate overnight or at least for several hours.

Just before you're ready to start cranking, remove the cold custard from the refrigerator and blend in the cold heavy cream. Pour the mixture into the cannister of the ice cream freezer and crank.

About vanilla powder: Vanilla beans are expensive (and hard to find in some places) but they're a real treat. To make the powder, grind several dried vanilla beans in a spice mill. One 4-inch bean will make about 2 teaspoons of powder. The tiny specks of vanilla will show, but in ice cream that's wonderful.

Some general hints about ice cream: Ice cream expands as it freezes. These recipes can be increased, but don't fill the container of the freezer more than three-fourths full.

Cream that's a day old makes a finer-grained product than fresh cream.

Preparing the mixture a day ahead of time makes a smoother product and also increases the yield.

CHEVON

Goat meat has a rather distinctive lamb-like flavor, and like lamb, it's somewhat dry and quite lean, so any recipes for lamb are equally good with chevon. Of course, neither requires any special recipes, especially for the creative cook. Depending on the cut, chevon can be fried, roasted, baked, broiled, grilled, or used in any recipes calling for ground meat, and of course made into sausage.

Dairy goats are not meat animals, and they don't provide as much meat as lambs. In addition, in most cases the most common or plentiful supply of chevon comes from young buck kids. The chops, for example, are very small, and we generally just leave them with the ribs. Barbequed ribs, then, are popular.

Barbequed Ribs

Barbeque Sauce:

1/2 cup chopped onion	1 clove garlic, crushed

Sauté together until tender in 1 tablespoon drippings.

1/2 cup water	2 tablespoons brown sugar
1 tablespoon vinegar	1/2 teaspoon salt
1 tablespoon Worcestershire sauce	1/4 teaspoon paprika
1/4 cup lemon juice	Optional: thyme, oregano, dry
1 cup chili sauce	mustard, hot peppers, or pepper
(or home-seasoned tomato sauce)	sauce

Simmer for 20 minutes.

Brown the ribs. Add the sauce and bake or grill until tender and done.

Marinated Chevon

Any meat can be marinated. This tenderizes it and often enhances the flavor. There are many marinades. Here's one suggestion:
(For about 3 pounds of chevon steak)

1/4 cup honey
1 large onion, sliced
2 teaspoons salt
1/4 teaspoon paprika
4 whole cloves
2 bay leaves

1 teaspoon oregano
1/2 teaspoon dry mustard
1 cup thick sour cream
cider vinegar and water
    (half and half) to cover

Arrange the steaks (or other meat) in a glass dish. Cover with water and vinegar. Add everything else except the cream. Set in the refrigerator for 2 days.

Then remove the meat and dry it. Dredge in flour and brown in butter. Add the marinade and simmer gently until tender.

Stir in the sour cream just before serving.

SOAP

Goat Milk Soap

A very nice soap can be made with goat milk.

Pour 1 can of lye into 3 pints of goat milk. (Be sure to follow the directions and observe the warnings on the lye can label.) Stir with a wooden spoon.

When the mixture is warm (don't touch the lye solution: just feel the outside of the container), pour 5-1/2 pounds of clear lukewarm fat into it (goat fat is best). Keep stirring while adding the fat.

Add 4 heaping teaspoons of Borax and 2 cups of finely ground oatmeal. Add 2 ounces of glycerine and stir 15-30 minutes or until it starts to harden.

Pour the mixture into molds (plastic foam drinking cups make fine molds) or, using rubber gloves, shape the soap into balls.

Let ripen for 3 weeks or more: older soap is better. It might have a peculiar odor at first, but this will disappear with age.

Where Milk Comes From

What makes a goat "let down" her milk? Why does she sometimes "hold back" milk? What acounts for the lactation curve, and what makes a goat dry off?

In brief, where does milk come from?

As a goat owner you're almost certain to ponder these questions while milking, sooner or later. Here are some of the answers.

Like all mammals, female goats produce milk for the purpose of feeding their offspring. Thus, any doe must be bred before she will lactate, and lactation stops naturally as kids are weaned. This period of milk production can be extended somewhat, but not indefinitely. Eventually the doe must be bred before she will start producing milk again.

ACTIVITY IN THE UDDER

The goat has two mammary glands, which collectively are called the udder.

Probably less than 50 percent of the milk an animal produces can be contained in the natural storage area of the udder: the balance is accommodated only by the stretching of the udder. In some cases this can cause the ligaments that suspend the udder to become permanently lengthened, causing the udder to "break away" from the body, resulting in what is called a pendulous udder.

A good udder is capacious, it has a relatively level floor, and is strongly attached. It has plenty of glandular tissue, but a minimum of connective and almost no fatty tissue. After milking, the normal, high-quality udder feels soft and pliable, with no lumps or knots which would indicate connective tissue resulting from injury or disease.

The udder is divided into right and left halves by a heavy membrane. The milk produced in each half can be removed only from the teat of that half.

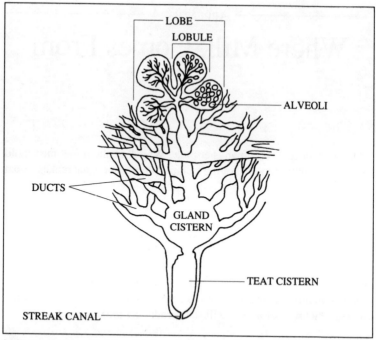

Milk travels from the alveoli, through the ducts, to the gland cistern and teat cistern, and out the streak canal.

THE TEAT, LOBES, AND ALVEOLI

Let's take a closer look at the mammary system, starting with the teat and working back to the origin of the milk itself.

Why are some goats hard to milk, while others are "leakers"?

The **teat** has an opening at the end, known as the **streak canal.** The streak canal is surrounded by sphincter muscles which prevent the milk from flowing out, and which determine how hard or easy it is to milk the animal. The strength of these sphincter muscles varies among individuals: those with very strong sphincters might be hard to milk, while those with weak sphincters can actually drip milk. (We'll see that these sphincters are affected by hormones resulting from premilking routines and other factors.)

Following the path of the milk back to its origins, we see where the teat widens into its **cistern,** the final temporary storage area before milking or suckling. Beyond that is the **gland cistern,** which is in the udder, and beyond that are large ducts that branch through all parts of the udder to collect and transport the milk toward the teat.

Each duct drains a single **lobe,** which further branches into **lobules,** which is where milk production actually takes place.

These lobules are composed of many tiny, hollow spheres called **alveoli,** which are said to resemble bunches of grapes. Very tiny grapes: one cubic centimeter holds about 60,000 of them. Milk is secreted in the cells of each alveolus.

Each alveolus is surrounded by muscle fibers, which balloon out as milk is secreted. When the animal is properly stimulated to let down her milk, these microscopic muscles contract, forcing out the milk.

BLOOD SUPPLY

Blood supply to the udder is important in the making of milk. For each unit of milk secreted, anywhere from 300 to 500 times as much blood passes through the udder!

Blood enters through the base of each half of the udder through the pudic artery. It returns to the heart for oxygenation via one of three passages, one of which is the subcutaneous abdominal vein, often called the milk vein. Many milkers believe the size and prominence of the milk vein is an indication of milking ability; however, experimenters have tied off these veins, and milk production wasn't appreciably affected.

Milk is made largely from constituents of blood. A tiny network of **capillaries** surrounds each alveolus. These capillaries carry blood to the base of the cells that line the alveolus. Milk is made when materials in the blood passing through these capillaries are taken up by the milk-making cells. The process involves, partly, filtration of certain constituents in the blood itself, and partly synthesis of constituents by cellular metabolism.

Scientists have been able to determine the blood constituents used in milk production by examining blood before it enters and after it leaves the udder.

Water—the largest component of milk (87 percent by weight)—is filtered from the blood. The vitamins and minerals of milk are filtered from the blood. The lactose in milk (its principle carbohydrate) is synthesized from blood glucose.

On the other hand, about 75 percent of the fat in milk is synthesized in the mammary gland. Most of this comes from acetate—which explains why animals on high-grain and low-forage diets often produce milk with a lower fat content: this diet results in reduced production of acetate in the rumen.

SECRETION

Now it's milking time. The cells of the alveoli have been busy making milk by filtering blood and synthesizing other constituents. These tiny cells lengthen as the milk accumulates. When filled, the cells rupture, pouring their contents into the lumen, the cavity of the alveolus. This causes the mammary glands to become saturated with milk, like a sponge.

Incidentally, all this activity is most rapid immediately *after* milking. As milk is secreted in the cells and collected, increasing pressure in the mammary system slows the secretion-discharge cycles. Each hour after milking, milk production decreases by 90 to 95 percent. At a certain point—technically, at a pressure of 30-40 millimeters in mercury, which is roughly equal to capillary pressure and about one-fourth of systemic blood pressure—milk secretion is reduced appreciably, or stopped entirely. If the animal isn't milked, the milk starts to be resorbed into the bloodstream.

Also of practical interest, when the udder is full the secreting cells are unable to rupture because of the pressure. Therefore, only that part of the milk that can pass through the semipermeable cell membrane can be discharged. The milk fat is not discharged. This lower-fat milk dilutes the milk previously secreted. That's why when the interval between milkings lengthens, the fat content of the milk decreases.

This also explains several other important phenomena, such as why milking three times a day results in more milk than two-time milkings: since secretion is highest soon after milking, with three milkings it progresses at a faster rate for more hours of each day. This is also the reason the strippings are higher in fat than the first milk drawn: the cells with accumulated milk fat are able to discharge the fat globules when udder pressure has been reduced. And in addition, it helps explain why the milk-fat percentage is usually higher with lower-producing animals or those whose production is declining in late lactation.

Now the stage is set: the goat is almost ready for you to start milking. Almost...but not quite! Coaxing milk from the udder involves more than just squeezing teats. What are required now are the only stimulators of lactation: hormones.

HORMONES

Hormones are closely related to milk production in many ways, one of the most obvious being the development of the udder itself. The udder is, after all, a secondary sex characteristic. Its very existence and function is the result of hormonal activity.

Six hormones are important in the intensity of lactation.

Prolactin. This proteohormone is secreted by the anterior pituitary gland and, in mammals, stimulates the initiation of lactation. It also increases the activity of the enzymes essential to the work of the epithelial cells (in the alveoli), which convert blood constituents to milk.

Thyroxine. Cows experimentally deprived of this product of the thyroid gland have gone down in milk production by as much as 75 percent.

Other research has shown that thyroxine secretion increases in the fall and winter, and decreases in spring and summer. This is said to partially explain why milk production decreases in hot weather.

Somatotropin. Today everyone should be familiar with this one: it's also known as bovine growth hormone, or BGH. This is also secreted by the anterior pituitary gland. It regulates growth in young animals, but also influences milk secretion by increasing the availability of blood amino acids, fats, and sugars for use by the mammary gland cells in milk synthesis.

Parathyroid hormone. This regulates blood levels of calcium and phosphorus, which are major constituents of milk. As such, it also plays a role in milk fever (parturient paresis). When an animal freshens and starts producing milk, the mammary glands rapidly withdraw calcium and phosphorus from the blood. Without proper nutrition—and parathyroid hormone—the animal can develop milk fever. The practice of feeding high levels of vitamin D before freshening is related to this hormone.

Adrenals. These hormones work both ways: small amounts are essential to milk production, but larger amounts will depress it. This is why it's important not to startle or frighten dairy animals. When disturbed, adrenaline is secreted to overcome the stress of the moment, but it decreases milk secretion.

Ocytocin. This is secreted by the hypothalamus and works with prolactin. (It also induces expulsion of the egg in the hen and is used to induce active labor in women or to cause contraction of the uterus after delivery of the placenta.)

STIMULI

A goat doesn't voluntarily "hold back" her milk. But she does have to be properly stimulated.

When she is, milk is suddenly expelled from the alveoli into the large ducts and udder cisterns. Then, and only then, can it be removed.

The natural stimulus is nursing. However, manual massage of the teats and udder (performed while washing them) has the same effect. In addition, sights, sounds, and smells have an effect on milk letdown. Your arrival at the barn, turning on the lights, feeding...all the routines of milking are signals for the milk to start pumping. This is one of the reasons many goats will produce less when moved to a new home.

Then ocytocin is poured into the bloodstream, reaching the udder in thirty to forty seconds. This causes the cells to contract, squeezing milk from the alveoli. Milk pressure in the cistern is almost doubled.

However, this lasts for only eight to twelve minutes. This backs up the belief that to get the most milk, you have to milk fast.

Ocytocin can be neutralized by adrenaline, which increases blood pressure, heart rate, and cardiac output. It also causes the tiny arteries and capillaries of the udder to constrict, and thus prevents ocytocin in the blood from reaching its destination. And adrenaline remains in the blood longer than ocytocin.

Adrenaline, of course, is released when the animal is in pain, frightened, irritated, startled by a loud noise, or otherwise bothered. (With females, embarrassment can also be added to this list.) Small wonder the first-time milker who pains, frightens, irritates, startles—and probably embarrasses the goat—gets so little milk!

DRYING OFF

You never get the "last drop" of milk from an udder. Normally, 10 to 25 percent remains even after stripping.

But as mentioned, if the milk isn't removed from the udder, the milk already there will be resorbed and the cells will quit producing more. Even just incomplete milking causes pressure to build more rapidly, and less milk is secreted between milkings. Eventually the secretion is impaired, and the animal dries off.

A goat can be dried off by milking on alternate days, or by stopping milking altogether.

OTHER FACTORS AFFECTING LACTATION

As we've seen, hormones play a major role in milk production. But they aren't the only factors. There are many others.

The first is genetics. If the animal hasn't inherited the potential to produce milk—including the capability to produce the needed enzymes—she won't produce milk.

The animal must have enough secreting tissue. A small udder has proportionately less of this tissue, and it also has less capacity for storing the milk secreted.

The stage of lactation plays a key role. Milk production generally peaks within a month or two of freshening, then drops off. The rate of this drop-off, or of persistence, has a marked effect on annual production. Persistent milkers might drop two or three percent a month; others drop off much faster. (Easy milkers, those that milk out more rapidly, are usually more persistent, for reasons already given.)

Frequent milking, also discussed earlier, tends to lengthen lactations.

Females in later stages of pregnancy generally drop off in milk production quite rapidly because nutrients are diverted from the mammary gland to the uterus, for growth and maintenance of the fetus.

Milk production increases with age, up to a point, but then drops off with advancing years. Milk production drops when an animal is in heat.

As might be expected, any disease can reduce milk production. Diseases can slow the circulation of blood to the udder, and that affects milk secretion.

High temperatures decrease milk production because they induce depressed appetites, reduced thyroid secretion, and other complications. Optimum temperatures seem to be between 50 and 80 degrees F.

And then of course there's feed. In the hierarchy that nature has wisely established, a body's first responsibility is survival. Maintenance comes before milk production.

Finally, there is your milking routine. If you don't properly stimulate your goat, you won't get all the milk possible, and that in turn will decrease production at future milkings.

Milking isn't just a chore: it's participating in a marvel of nature. And lucky people who have goats get to do it twice a day!

The Composition of Milk

"Goat milk is richer than cow milk, isn't it?"

If we assume "richer" means higher in fat, the answer is "sometimes." There is a great deal of variation in the composition of milk, not only among species but also among breeds, families, and even the same individuals at different ages and stages of lactation, or on different feeds.

If you look at one of those charts giving the "average" composition of milk (and seldom do two of them agree), here's what you might find:

Mammal	Fat	Protein	Lactose	Minerals	Total Solids
Human	3.7%	1.6%	7.0%	0.2%	12.5%
Cow	4.0	3.3	5.0	0.7	13.0
Goat	4.1	3.7	4.2	0.8	12.9

Similar charts for various *breeds* of cows show greater differences than those indicated between cows and goats. The same is true for goats. While *averages* show that Nubians produce milk that's richer in fat than Saanens, these too are only averages. Some Saanens produce milk that's richer than the milk of some Nubians.

Some of the reasons are explained in Appendix A (i.e., diet, intervals between milkings). In addition, milk composition is highly heritable. Animals of the same bloodlines are likely to have the same milk characteristics (given the same health, feed, age, state of lactation, and other factors).

The milk of individual animals changes drastically at different periods of time, too. The most dramatic example is colostrum.

COMPOSITION OF COLOSTRUM AND NORMAL MILK
(COW, AVERAGE PERCENT)

Component	Colostrum	Normal Milk
Water	71.7%	87.0%
Milk fat	3.4	4.0
Casein	4.8	2.5
Globulin & albumin	15.8	0.8
Lactose	2.5	5.0
Minerals	1.8	0.7
Total solids	28.3%	13.0%

While colostrum presents an extreme example, changes in milk composition do occur throughout lactation, with milk produced during the first two months after freshening generally being from 0.5 to 1.5 percent lower in fat than milk from the same animal during the last two months of lactation.

Protein, fat, and SNF (solids not fat) generally decline with age, with SNF declining even more than fat.

Butterfat tests normally rise in the fall and drop in spring.

Other factors that can cause minor changes in the composition of milk include temperature (temperatures above 70°F. and below 30°F. increase the fat content) and exercise (slight exercise slightly increases fat content, while more strenuous exercise decreases fat and total output).

Perhaps the greatest single reason the myth of rich goat milk persists, however, is that few people today have ever tasted milk that has not been standardized. Milk sold in stores has fat removed to meet minimum government standards, and of course, the tasteless chalk-water called 2 percent and even 1 percent have become common. The fact is, when that milk left the cow it was probably just as rich as goat milk.

Injections

A reader asks, "How do I give a goat an injection?"

The best way to learn is by watching someone else. Most veterinarians will show you how to do something as basic as this: they have more important things to do with their time and education than give shots. But here is some general information.

Injections can be intramuscular (IM), subcutaneous (SC), intraperitoneal (IP), and intravenous (IV). These names refer to where the needle goes in. Intramuscular means within the muscle, or meat; generally this means a rear leg. Subcutaneous means under the skin. Intraperitoneal means within the peritoneum, or the space between the intestines and other parts of the lower gut and the internal organs such as the liver. Intravenous means directly into a vein (this one is best left to a vet).

Different medications require different types of injections. Read all labels carefully, and not only to learn how the medication should be administered. Note also any safety precautions, how the medicine should be stored, shelf life, withdrawal time, and any other information provided.

There are three commonly used types of syringes: all glass, glass and metal, and plastic. Most farms use disposable plastic syringes, as these don't require sterilization equipment. (Note: equipment can be sterilized in a domestic pressure cooker by placing them in steam under pressure for thirty minutes. Remove metal plungers so they do not expand and break the syringe. Holding it in a pan of boiling water for thirty minutes is another method, but not totally safe.) Disposable syringes come in sterile packages, are used once, and discarded.

There are also reusable needles, which require sterilization, and disposable needles. These come in various lengths and gauges. One-inch

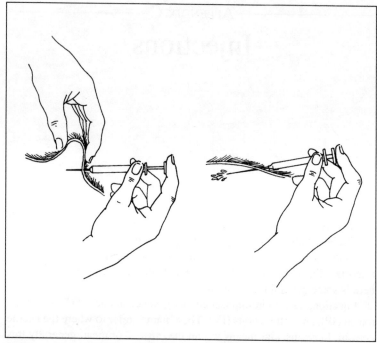

To give a sub-cu injection, (a) pull up loose skin to make a "tent"; insert needle, avoiding muscle or joints. (b) Let go of the skin when needle is in and push plunger to inject medication under the skin.

needles in 16-, 18-, and 20-gauge sizes are commonly used for goats. The larger the gauge the smaller the needle: the choice usually depends on the medication being given, although this is largely a matter of personal preference.

There are several ways to fill the syringe. Here is one method.

1. Wipe the rubber bottle cap with alcohol or disinfectant.

2. Place a sterile needle on a sterile syringe, pull back the plunger, and fill the syringe with air.

3. Insert the needle through the rubber cap, invert the bottle, and blow a little air into the bottle by depressing the plunger. Repeat blowing in air and withdrawing fluid until the syringe is full.

Depress the plunger until the syringe holds the desired amount.

It's not possible to *sterilize* the skin of a live animal, but it should be as clean as possible. A wide variety of skin disinfectants are available, including 70 percent solution of ethyl alcohol in water, and iodine, including tincture of iodine and Lugol's iodine.

Most vaccines and antisera are given by the subcutaneous route, or SC. The needle is inserted through the skin, and the substance is deposited beneath the skin.

Grasp the skin between the thumb and forefinger and raise it into a "tent" or bubble. Insert the needle at the base of this tent, parallel to the skin. Let go of the skin when the needle is in. Push the plunger of the syringe to force the substance under the skin. Withdraw the needle and syringe.

Follow the manufacturer's instructions exactly when using and storing vaccines and medications.

Immunizations

Every region has different health problems that can be prevented by immunizations. The person best qualified to help you set up and administer a health maintenance program is your veterinarian.

Contacting a vet about vaccinations has several side benefits. It establishes contact: the doctor will know you have goats and that you care enough about them to seek professional guidance before some dire emergency arises. And a vet is more likely to proffer valuable advice or assistance on other matters on a routine, nonemergency call than in a life-or-death situation.

Chances are your vet will recommend the four basic vaccinations: tetanus, white muscle disease, enterotoxemia, and pasteurellosis.

(While the most prevalent goat ailment is probably caseous lymphadenitis, and the AIDS-like CAE gets the most publicity these days, there are no vaccines for either of these. Nor are there vaccines for other viral-caused dieseases, such as pneumonia and coccidiosis.)

There are several things to know about vaccines.

Remember that a vaccine is not medicine, in the sense of being a "cure." It's a preventive measure. If your goat has already come down with enterotoxemia, it's too late for a vaccination.

Also keep in mind that vaccines often differ according to their manufacturer. No book can give blanket instructions regarding these products, but even more importantly, just becuase you've used a vaccine once, don't assume it will be the same next time. *Always* read and follow instructions on the labels carefully.

With these caveats, here are a few guidelines.

Enterotoxemia: Five cc. sub- cu (under the skin) to pregnant does two weeks before freshening. Kids get 2-1/2 cc. at four months of age.

White muscle disease: 2-1/2 cc. Bo-Se per 100 pounds of body weight to pregnant does (sub-cu) one week before freshening.

Tetanus: 1/2 cc. IM (intramuscular) four weeks before freshening. This will protect young kids for disbudding and castration, but they should get a booster shot at two months of age.

Pasteurella: Two cc. IM to kids at two months of age. Repeat in two weeks.

There are others, soremouth being one of the more prominent. But many goat owners resist vaccinating kids for soremouth: it gives kids a sometimes bad case of the disease.

But of course there are always dangers. All injections bring the risk of anaphylactic reaction. (Anaphylaxis is sensitivity to drugs or foreign proteins introduced into the body resulting from sensitization following prior contact with the causative agent.) Always watch your animals for ten minutes after any injections, and have a bottle of epinephrine on hand, just in case. And keep your vet's phone number handy.

Glossary

Abomasum: The fourth or true stomach of a ruminant where enzymatic digestion occurs.

Abscess: A boil; a localized collection of pus.

ADGA: American Dairy Goat Association, the oldest and largest dairy goat registry in the United States.

Afterbirth: The placenta and associated membranes expelled from the uterus after kidding.

AGS: American Goat Society, a registry.

AI: Artificial insemination.

American: A doe that is 7/8th purebred and recorded with ADGA; a buck that is 15/16ths purebred and recorded with ADGA.

Anthelmintic: A drug that kills worms.

AR (advanced registry): A designation indicating a goat has produced at least 1,500 pounds of milk in a 305-day lactation.

Ash: The mineral matter of a feed; what is left after complete incineration of the organic matter.

Balling gun: Device used to administer a bolus (a large pill).

Barn records: A tally of daily milk production kept by the goat owner rather than by an official testing organization.

Blind teat: A nonfunctioning half of an udder (usually due to mastitis).

Bloat: An excessive accumulation of gas in the rumen and reticulum resulting in distension.

Bolus: A large pill for animals; regurgitated food that has been chewed (cud).

Breed: Animals with similar characteristics of conformation and color, which when mated together produce offspring with the same characteristics; the mating of animals.

Breeding season: The period when goats will breed, usually from September to January.

Buck: A male goat. Buckling: A young male.

Browse: Bushy or woody plants; to eat such plants.

Buck rag: A cloth rubbed on a buck and imbued with his odor and kept in a closed container; used by exposing to a doe and observing her reaction to help determine if she's in heat.

Burdizzo: A device used to crush the spermatic cords to render a buck or buckling sterile.

Butterfat: The natural fat in milk; cream.

CAE (caprine arthritis encephalitis): A serious and widespread type of arthritis, caused by a retrovirus.

California mastitis test (CMT): A do-it-yourself kit to determine if a doe has mastitis.

Caprine: Pertaining to or derived from a goat.

Carbonaceous hay: Any hay that is not a legume, including timothy, brome, Johnson grass, and Bermuda grass.

Chevon: Goat meat.

Classification: A system of scoring goats based on appearance.

Colostrum: The first thick, yellowish milk a goat produces after giving birth, rich in antibodies without which the newborn has little chance of survival.

Concentrate: The nonforage part of a goat's diet, principally grain but including oil meal and other feed supplements, which is high in energy and low in crude fiber.

Conformation: The overall physical attributes of an animal; its shape and design.

Creep feeder: An enclosed feeder for supplementing the ration of kids, but which excludes larger animals.

Cull: To remove a substandard animal from a herd; a substandard animal itself.

Dairy cleaning agents: Alkaline or acid detergents for washing milking equipment; iodine or chlorine compounds for sanitizing milking equipment.

Dam: Female parent.

DHIA (Dairy Herd Improvement Association): A program administered by the USDA through Extension Services to test and record milk production of cows and goats.

DHIR (Dairy Herd Improvement Registry): A milk-production testing program administered by dairy goat registries in cooperation with DHIA.

Disbudding iron: A tool, usually electric, that is heated to burn the horn buds from young animals to prevent horn growth.

Dish face: The concave nose of the Saanen.

Doe: A female goat. Doeling: A young female.

Dry period: The time when a goat is not producing milk.

Drylot: An animal enclosure having no vegetation.

Elastrator: A device used to apply a rubber band to the scrotum so it will atrophy and fall off.

Electrolyte: Mineral salts necessary for life, including sodium, potassium, calcium, and magnesium, which are lost when a body loses more fluid than it can take in.

Feed additive: Anything added to a feed, including preservatives, growth promotants, and medications.

Flushing: Feeding females more generously two to three weeks before breeding in order to stimulate the onset of heat and induce the shedding of more eggs. This results in more offspring and improves the chances of conception.

Forage: The hay or grassy portion of a goat's diet.

Free choice: Free to eat at will with food (esp. hay) always present.

Freshen: To give birth (kid) and come into milk.

Gestation: The time between breeding and kidding (average 150 days).

Grade: A goat that is not purebred, or cannot be proven pure by registry records; any goat of mixed or unknown ancestry.

Grade A: A category of licensed dairy that meets strict regulations for equipment, milk handling, and sanitation.

Growthy: Description of an animal that is large and well developed for his age.

Hand feeding: Providing a measured amount of feed at set intervals.

Hand mating: Controlled breeding, as opposed to letting the male run loose with or in a pen of unbred females.

Hay: Dried forage.

Heat: Estrus; the condition of a doe being ready to breed.

Hermaphrodite: A sterile animal with reproductive organs of both sexes, generally associated with the mating of two naturally polled animals.

Homozygous: Containing either, but not both, members of a pair of alleles.

Hormone: A chemical secreted into the bloodstream by an endocrine gland, bringing about a physiological response in another part of the body.

Horn bud: Small bump from which horns grow.

Intradermal: Into or between the layers of the skin.

Intraperitoneal: Within the peritoneal cavity.

Intravenous: Within a vein.

IM: Intramuscular (within the muscle).

Inbreeding: The mating of closely related individuals.

IU: International unit, a standard unit of potency of a biologic agent such as a vitamin or antibiotic.

Kid: A goat under one year of age; to give birth.

Lactation: The period in which a goat is producing milk; the secretion or formation of milk.

Lactation curve: Daily milk production as represented on a graph, usually showing a rapid rise soon after freshening, then a slow fall.

Legume: A family of plants having nodules on the roots bearing nitrogen-fixing bacteria, including alfalfa and the clovers.

Linear appraisal: A system of scoring goats on individual conformation traits.

Linebreeding: A form of inbreeding that attempts to concentrate the genetic makeup of some ancestors.

Mastitis: Inflammation of the udder, usually caused by an infection.

Microorganism: Any living creature of microscopic size, especially bacteria and protozoa.

Milking bench (or stand): A raised platform that a goat stands upon to be milked, usually with a seat for the milker and a stanchion for the goat's neck.

Milking through: Milking a goat for more than one year.

New Zealand fencing: A system of electric fencing using a high-energy charger.

Off feed: Not eating as much as normal.

Out of: Mothered by.

Overconditioned: Overfed; fat.

Papers: Certificates of registration or recordation.

Pedigree: A paper showing an animal's forebears.

Polled: Naturally hornless.

Precocious milker: A goat that produces milk without being bred.

Protein supplement: A feed product containing more than 20 percent protein.

Purebred: An animal whose ancestry can be traced back to the establishment of a breed through the records of a registry association.

Raw milk: Milk as it comes from the goat; unpasteurized milk.

Recorded grade: A goat, either not purebred or not verifiably purebred, that is recorded with ADGA.

Registered: A goat whose birth and ancestry is recorded by a registry association.

Rennet: An enzyme used to curdle milk and make cheese.

Retained placenta: A placenta not expelled at kidding or shortly thereafter.

Reticulum: The second compartment of the ruminant stomach.

Roughage: High fiber, low total digestible nutrient feed, consisting of coarse and bulky plants or plant parts; dry or green feed with over 18 percent crude fiber.

Rumen: The first large compartment of the stomach of a goat where cellulose is broken down.

Scours: Perisitent diarrhea in young animals.

Service: Mating.

Settled: Having become pregnant.

Sire: Male parent; to father.

Stanchion: A device for restraining a goat by the neck for feeding or milking.

Standing heat: The period during which a doe will accept a buck for mating, usually about 24 hours.

Star milker: A designation of high milk production based on a one-day test, not the entire lactation. *M, **M, etc., indicates that the dam and granddam also hald *M status. *B or star buck indicates star milkers in a buck's family tree.

Straw: Dried plant matter (usually oat, wheat, or barley leaves and stems) used as bedding; the glass tube semen is stored in for AI.

Strip: To remove the last milk from the udder.

Strip cup: A cup into which the first squirt of milk from each teat is directed, used to show any abnormalities that might be in the milk.

Subcutaneous: Beneath the skin.

Synthesis: The union of two or more substances to form a new material.

Tattoo: Permanent identification of animals produced by placing indelible ink under the skin, generally in the ear but in the tail web of La Manchas.

Test (to be on test; official test): To have daily milk production weighed and its butterfat content determined by a person other than the goat's owner.

Total digestible nutrient (TDN): The energy value of a feed.

Trace mineral: A mineral nutrient essential to animal health, but used only in very minute quantities.

Type: The combination of characteristics that makes an animal suited for a specific purpose, such as "dairy type" or "meat type."

Udder: An encased group of mammary glands provided with a teat or nipple.

Udder wash: A dilute chemical solution, usually an iodine compound, for washing udders before milking.

Unrecorded grade: A grade goat not recorded with any registry association.

Upgrade: To improve the next generation by breeding a doe to a superior buck.

Vermifuge: Any chemical substance administered to an animal to kill internal parasitic worms.

Wattle: Small, fleshy appendage.

Wether: A castrated buck.

Whey: The liquid remaining when the curd is removed from curdled milk when making cheese.

WMT (Wisconsin Mastitis Test): A do-it-yourself kit to determine if a doe has mastits.

PHYSIOLOGICAL DATA

Temperature: Normal rectal temperature of the goat ranges from 101.5° to 104.0° F. This varies with the air temperature, exercise or excitement, and amount of hair. To determine if an individual goat's temperature is abnormal, compare it with several others under the same conditions.

Pulse: 70-80/min.

Respiration: 12-20/min.

Puberty: 4 to 12 months.

Estrus cycle: 18-23 days.

Length of heat period: Average 18-24 hours; range from 12-36 hours.

Gestation period: 148-153 days; average 150 days.

Birth weight: from 4.5-4.9 pounds.

Some Suggested Sources for Further Information

Publications
United Caprine News
P.O. Drawer A
Rotan, TX 79546
(Good monthly newspaper)

Dairy Goat Journal
6041 Manona Drive
Manona, WI 53716
(Monthly magazine heavy on showing)

Hall Press
P.O. Box 5375
San Bernardino, CA 92412
(Books on goats)

Statewide Goat Sales
P.O. Box 204
Wellington, OH 44090
(Monthly publication)

Small newsletters are issued by many of the local and regional goat clubs. Very often their location and quality vary from year to year as volunteer editors change, but some of them are quite good. Ask the people you buy goats from, other goat people, and your county agent about any such clubs in your area.

Some of the goat supply catalogs are very informative, particularly the one from Caprine supply—which also lists an excellent assortment of books.

Countryside & Small Stock Journal, W8333 Doepke Rd., Waterloo, WI 53594 has a regular dairy goat department edited by the author of this book. *Hoard's Dairyman,* 28 Milwaukee Ave. West, Ft. Atkinson, WI 53538 sometimes has goat articles and provides valuable general dairying information in its regular coverage of cows.

Pennsylvania State University, 307 Agricultural Administration Building, University Park, PA 16802 has a correspondence course on dairy goats.

Some state 4-H programs have literature on dairy goats.

Supplies
American Supply House
P.O. Box 1114
Columbia, MO 65205
(Goat supplies)

Caprine Supply
33001 West 83rd St.
DeSoto, KS 66018
(Goat supplies; very complete and
helpful catalog)

Cheesemaking Supply Outlet
260 Moore Rd.
Butler, PA 16001
(Cheesemaking supplies)

Dairy Goat Nutrition
P.O. Box 22363
Kansas City, MO 64113
(Feed supplements)

Hoegger Supply Company
160 Providence Road
Fayetteville, GA 30214-0991
(Goat and cheesemaking supplies)

Kritter Korner
Rt. 1
Broughton, IL 62817
(Hoof trimmers)

Lehman Hardware
P.O. Box 41
Kidron, OH 44636
(Dairy and cheesemaking supplies)

NASCO
901 Janesville Ave.
Ft. Atkinson, WI 52538
(General farm supplies)

New England Cheesemaking
Supply Company
Box 85
Ashfield, MA 01330
(Cheesemaking books and supplies)

Norseman Sausage Supplies
Rt. 2, Box 141-A
Wellsville, KS 66092
(Sausage-making supplies and
equipment)

Nutritional Research
Associates, Inc.
P.O. Box 354, 407 E. Broad St.
South Whitley, IN 46787
(Feed supplements)

Premier Fence Systems
Rt. 1
Washington, IA 52353
(Fencing)

Rick Skillman Welding
608 South 5th St.
Douglas, WY 82633
(Milking stands)

Waterford Corp.
P.O. Box 1513
Ft. Collins, CO 80522
(Fencing)

Registry Associations
American Dairy Goat Association
P.O. Box 865
Spindale, NC 28160

American Goat Society, Inc.
Rt. 1, Box 56
Esperanle, NY 12066-9704

International Dairy Goat Registry
Rt. 1, Box 265
Rossville, GA 30741

National Pygmy Goat Association
P.O. Box 380
Maple Valley, WA 98038

Breed Clubs
Alpines International
c/o Joan B. Godfrey
557 South St.
Suffield, CT 06078

American Angora Breeders
    Association
Rock Spring, TX 78880

American La Mancha Club
1940 Bumblebee Dr, NW
Chino Valley, AZ 86323

International Nubian Breeders
    Association
P.O. Box 130
Creswell, OR 97426

National Saanen Breeders
    Association
Mimi Waterman, Secy-Treas.
3 Kerr Rd.
Canterbury, CT 06331

National Toggenburg Club
Jan McMahin, Secy-Treas.
4044 Polaris
Joshua Tree, CA 92252

Nigerian Dwarf Goat Club
Rt. 1, Box 368
Red Rock, TX 78662

Oberhasli Breeders of America
Jo Ann Clugston, Secy-Treas.
RD 3, Box 145C
Manhein, PA 17545

Sable Breeders Association
2811 Michael Dr.
Newbury Park, CA 91320

Artificial Insemination
All American AI
Rt. 1, Box 60
Coquille, OR 97423

Buck Bank
2344 Butte Falls Hwy.
Eagle Point, OR 97524

Buckskin AI
1380 2nd St.
Anderson, CA 96007

Magnum Semen Works
2200 Albert Rill Rd.
Hampstead, MD 21074

Miscellaneous
Arian Dairy Goats
P.O. Box 70343
Denver, CO 80207
(Educational slides, ceramics, magnets)

Arkansas Gold
P.O. Box 210
Lamar, AR 72846
(Goat milk cheese gift packs)

Caprifield
24639 Garrish Valley Rd.
Yamhill, OR 97148
(Signs, ornaments, weather vanes)

Cosmic Software
P.O. Box 43551
Louisville, KY 40243
(Computer software)

Goat Pedigree
Box 30862
Tucson, AZ 85751
(Computer software)

Goats "R" Us
P.O. Box 478
Rogue River, OR 97537
(Caprine novelties)

Khimaira Kaprine Kreations
Rt. 2, Box 223
Luray, VA 22835
(Caprine novelty and gift-type items)

Mattox Farms
1380 Spry Rd.
Zanesville, OH 43701
(Broom covers)

Paan Goat Milk Soap
P.O. Box 85
Powhatan, VA 23139
(Goat milk soap)

Index

Other Storey/Garden Way Publishing Books You Will Enjoy

Raising Sheep the Modern Way, Revised and Updated Edition, by Paula Simmons. The small-scale sheep raiser's bible includes information on: new theories about breeding, pregnancy management, and handling of lambs; using guard dogs; medications and vaccines; plus all the housing, feeding, and general care information needed to get started with sheep. 272 pages, 6x9, illustrated quality paperback, $9.95. Order #529-4.

Raising Rabbits the Modern Way, Revised and Updated Edition, by Bob Bennett. The best book available on modern rabbit raising techniques, with complete coverage of building wire hutches, feeding, breeding, marketing, and all aspects of managing a small rabbitry. 192 pages, 6x9, illustrated quality paperback, $8.95. Order #479-4.

Raising Poultry the Modern Way, Revised and Updated Edition, by Leonard Mercia. Stock selection, feeding, brooding, management, and disease prevention for laying flocks, meat chickens, turkeys, ducks & geese. Written by an Extension Service expert especially for the small grower. 224 pages, 6x9, illustrated quality paperback, $8.95.

Raising the Home Duck Flock, by Dave Holderread. For persons who want to raise ducks but don't know how to get started, this book explains how many ducks you should have for various conditions, which breeds may be best for you, where and how to buy them. 192 pages, illustrated quality paperback, $7.95. Order #169-8.

Small-Scale Pig Raising, by Dirk van Loon. From the beginning, this book explains exactly why raising a feeder pig is the best bet for someone with little land who wants to produce the most meat for the smallest investment of time and money. 272 pages, 6x9, illustrated quality paperback, $9.95. Order #136-1.

Keeping Livestock Healthy, by N. Bruce Haynes, D.V.M. Updated Edition. A how-to-do-it veterinary guide. Covers all farm animals. *"A must."* —American Veterinary Journal. 324 pages, 6x9, illustrated quality paperback, $14.95. Order #409-3.

Tan Your Hide! Home Tanning Leathers & Furs, by Phyllis Hobson. Easy methods for the home tanner and fur skin worker. 140 pages, 6x9, illustrated quality paperback, $6.95. Order #101-9.

Basic Butchering of Livestock & Game, by John J. Mettler Jr., D.V.M. With this easy-to-read text you will find: 140 how-to drawings (very easy to follow every step); recommended tools and equipment; processing and preserving; meat inspection information. 206 pages, 6x9, illustrated quality paperback, $10.95. Order #391-7.

Raising a Calf for Beef, by Phyllis Hobson. All the information you need to raise a calf, with complete butchering instructions. 128 pages, 6x9, illustrated quality paperback, $6.95. Order #095-0.

The Canning, Freezing, Curing & Smoking of Meat, Fish & Game, by Wilbur F. Eastman, Jr. An authoritative work on the art of home processing of meat, fish, and game. Step-by-step instruction. 220 pages, 6x9, illustrated quality paperback, $6.95. Order #045-4.

Building Small Barns, Sheds & Shelters, by Monte Burch. Provides basic, easy-to-follow construction methods for attractive outbuildings. Plans for multi-purpose barns and barn-style garages, woodshed, toolshed, carport, and housing for poultry, rabbits, and hogs. 236 pages, 6x9, illustrated quality paperback, $12.95. Order #245-7.

These books are available at your bookstore, feedstore, lawn and garden center, or directly from Storey/Garden Way Publishing, Schoolhouse Road, Pownal, VT 05261. Send for our free mail-order catalog. Please send $2.75 per order for postage and handling.